Optoelectronics

ENDEL UIGA

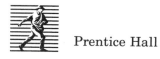

Prentice Hall

Englewood Cliffs, New Jersey   Columbus, Ohio

Library of Congress Cataloging-in-Publication Data

Uiga, Endel.
    Optoelectronics / Endel Uiga.
       p.   cm.
    Includes bibliographical references.
    ISBN 0-02-422170-8
    1. Optoelectronics.   I. Title.
  TA1750.U35   1995
  621.381'045—dc20
                                         94-31912
                                           CIP

Cover photo: Tony Stone Worldwide
Editor: Charles E. Stewart, Jr.
Production Editor: Julie Anderson Tober
Production Supervision: Beacon Graphics Corporation
Cover Designer: Thomas Mack
Production Buyer: Pamela D. Bennett
Illustrations: Academy Artworks

This book was set in Century Schoolbook by Beacon Graphics and was printed and bound by R. R. Donnelley and Sons Company. The cover was printed by Phoenix Color Corp.

© 1995 by Prentice-Hall, Inc.
A Simon & Schuster Company
Englewood Cliffs, New Jersey 07632

Photo credits: Hamamatsu Corporation, p. 222, 308; Hewlett Packard Corporation, p. 189, 196; Industrial Electronics Engineers, Inc., p. 94, 116, 124; Laser Diode Products, Inc., p. 156; Lee Laser, Inc., p. 148; NTC Laser Machine Group Corp., p. 161; Photo Research Corporation, p. 86, back cover; Planar Systems, Inc., p. 111; Rofin-Sinar, Inc., p. 151; Tektronix, Inc., p. 41.

Printed in the United States of America

10 9 8 7 6 5 4 3 2 1

ISBN: 0-02-422170-8

Prentice-Hall International (UK) Limited, *London*
Prentice-Hall of Australia Pty. Limited, *Sydney*
Prentice-Hall of Canada, Inc., *Toronto*
Prentice-Hall Hispanoamericana, S.A., *Mexico*
Prentice-Hall of India Private Limited, *New Delhi*
Prentice-Hall of Japan, Inc., *Tokyo*
Simon & Schuster Asia Pte. Ltd., *Singapore*
Editora Prentice-Hall do Brasil, Ltda., *Rio de Janeiro*

To Elise — for 54 good years

Preface

This book was born out of a dire need. Several years ago, at County College of Morris, I was given the task of organizing an optoelectronics course with supporting laboratory. Before starting my work, I sent a questionnaire to, and conducted several interviews with, Northern New Jersey electronics companies to determine their interests and needs in this subject. To my surprise, the interest was not only fiber optics technology, which I expected, but was evenly divided over the entire optoelectronics technology spectrum, including basic radiometric and photometric theory, optics, and a host of optoelectronics devices and their applications. I compiled my syllabus accordingly. There I ran into my first problem: I could not find a single technology or undergraduate level textbook that came close to my needs. Thus, I wrote a collection of handouts to complement the course, expanding them each year. After a few years my colleagues suggested that I write a book, based on these notes. I foolishly agreed, not realizing how much work it would require to turn a collection of notes into an organized book. But here it is, after several years of writing pains and the help of many friends and colleagues.

Both at heart and by lifelong experience, I am an electronics designer. Thus, this book is a design-oriented text. Ability to design an electronics gadget that works reliably, according to set specifications, is the ultimate proof of one's understanding of theory and applications. A designer's interest, contrary to a scientist's, is not to poke deeply into inner workings of devices but to make them work predictably and reliably. This book introduces the principles of optoelectronics design to students or engineers who have some background in general electronics design, circuit theory, electronics devices, and digital techniques. In addition, this book adds radiometric and photometric theory, basic optics, and an introduction to optoelectronics devices and methods to their toolbox.

The subject matter is approached in a most simple, design-oriented way. Three steps are used to introduce the optoelectronics devices: (1) their characteristics are defined and explained; (2) they are classified among similar devices; (3) design rules, applications and limitations are introduced. Operating principles of devices are presented only in a very simplified and elementary manner since in most cases they are not essential for device applications.

The devices that are covered are optoelectronics sources, detectors, lasers and laser diodes, displays, optocouplers, and fiber optics devices. Emphasis is placed on devices that are usable as components in optoelectronics designs. More complex devices, such as complex laser systems and complicated large-scale displays, are described only in general terms. One exception is fiber optics. This subject is worth a good semester's course by itself and since good books for this technology are available, only a survey of available devices and methods is given.

The book definitely contains more material than can be fitted into a one-semester course. Thus, the instructor may make a selection of subject matter that is in his or her sphere of interest. For the student, and for an engineer entering optoelectronics, the book may be a valuable self-study source for solving a variety of optoelectronics problems that must be faced in a professional career.

Since I am an engineer and not a writer, and English is not my first language, I have to thank many friends and colleagues who helped to bring this book to its conclusion. Among them are my sons-in-law Andres Värnik, an accomplished electronics engineer, Dr. Herb Laeger, a recognized expert in the field of lasers, and John Hughes, an English teacher. Special appreciation is extended to my colleague and long-time friend, Walley White, who read the manuscript and gave many valuable suggestions. My former student, Rodger Stahl, was very helpful in checking my arithmetic (and discovering many errors in the process). Mr. Stuart Kenter, a professional editor, however, deserves most of the credit for transforming my engineer's lingo manuscript to a readable text. Apart from help in technical fields, I have to thank my dear wife, Elise, for moral support. Over many years she never got tired of, or complained about, my perennial excuse for avoiding my household chores: "Sorry, I have to work on my book."

Endel Uiga

Contents

Optoelectronics

1 Radiometry and Photometry

1-1 THE AWARENESS OF LIGHT

Early peoples realized that life on our planet depends on light and radiation from the sun. In fact, many ancient cultures granted the sun the status of a god. Sun cycles were identified and memorized, and the solstices of winter and summer were commemorated with fire and ritualistic celebrations. Given this long-standing awareness of the influence of light on life, it is somewhat surprising that serious attempts to explain the nature of light came relatively late in history. It was not until the middle of the seventeenth century, during a period of scientific awakening, that two theories concerning the nature of light emerged, almost simultaneously.

Early Theories of Light

Newton's Theory.   One of the first comprehensive theories of light was advanced by the famous English scientist *Sir Isaac Newton* (1642–1727), who postulated that a light ray consists of a stream of fast-moving particles called **corpuscles.** Newton believed that corpuscles emitted from a light source travel in straight lines and that the corpuscles can be blocked by opaque objects (see Figure 1–1). Corpuscles, in Newton's theory, reflect from smooth surfaces in the same manner as a ball bounces off concrete, spread their energy, and provide illumination in inverse proportion to the surface area they strike. Newton's basic concept follows our everyday experience of light and is analogous to the laws of mechanics.

To explain color, Newton proposed that corpuscles come in a variety of colors which, in turn, combine to form additional colors. He maintained that

FIGURE 1–1   A simplified illustration of three theories of light mentioned in the text. The image of a pinhole according to each of the theories.

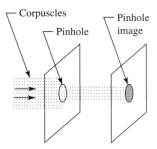

(**a**) Pinhole image according to Newton's theory

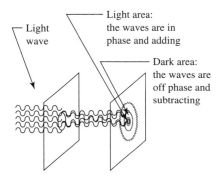

(**b**) Pinhole image according to wave theory

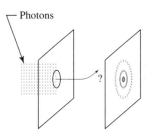

(**c**) Pinhole image according to quantum theory

types of corpuscles have differing refractive properties when they pass through transparent materials, thus explaining why rainbow colors appear when light passes through a prism.

Newton's hypotheses, presented in a treatise entitled *Opticks*, interpreted most common optical phenomena. It failed, however, to explain diffraction and polarization.

Huygens' Theory.   A different theory on the nature of light was formulated by Dutch scientist *Christiaan Huygens* (1629–1695), who proposed that light emitted from a source is actually a wavelike disturbance transmitted through the hypothetical medium called **ether,** a substance that was thought to fill the entire universe. Huygens reasoned that since Newton's corpuscular rays are made of particles, they must of necessity have to exhibit some form of interference when they intersect. Such was not the case. Waves, on the other hand, could cross each other without effect.

Support for the Wave Theory

Because of the influence of Newton's preeminent authority and the fact that his ideas about light conformed to recognized mechanical principles, his corpuscular theory of light was accepted for some time. At the end of the eighteenth century, however, the results of certain experiments conducted by the Englishman *Thomas Young* (1773–1829) gave strong support to the wave theory of Huygens. When Young projected light through a pinhole and onto a screen, he observed not just a single bright spot but, rather, a pattern of many brighter and darker concentric circles surrounding the spot. Young explained this pattern—which we now call an **interference pattern**—by the wave theory of light [see Figure 1–1(b)]. From this experiment, Young was able to calculate wavelength, which led him to discover that wavelengths were different for each color of light. Moreover, his experiments demonstrated that light waves are transverse waves; that is, the oscillation is perpendicular to the wave's direction of propagation. This discovery accounted for the phenomenon of polarization as well. (see p. 2)

Even stronger support for the concept that light consisted of waves came from British physicist *James Clerk Maxwell* (1831–1879), who in the middle of the nineteenth century developed a unified electromagnetic theory. He predicted the presence of electromagnetic waves that travel at the same speed as light. Maxwell's theory, together with conclusions drawn from the results of several other experiments at the end of the nineteenth century, convinced scientists that light was indeed an electromagnetic wave. Furthermore, the wave theory explained the behavior of light so well that it was given the same credence accorded to the theory of gravitation, which accounted for the movement of the planets. Nonetheless, a few minor discrepancies demanding solutions lingered in the wave theory.

Further Discoveries about Light

The Quantum.   One of these "minor discrepancies" involved the fact that the frequency spectrum of the radiation from a heated body—known in physics as **black body radiation**—did not agree with the calculations. Working with this problem, German physicist *Max Planck* (1858–1947) developed, at the turn of the twentieth century, an exact equation for the radiated spectrum. Planck boldly assumed that electromagnetic energy occurred in small, discrete chunks rather than in continuous form. The idea that energy exists in discontinuous chunks was revolutionary, completely foreign to the classical theories of physics. Planck named the chunks of energy **quanta,** assigning the following frequency-dependent value to the quantum:

$$E = hf \tag{1-1}$$

where E = energy of quantum (J)
$\quad$ h = Planck's constant 6.62×10^{-34} (Js)
$\quad$ f = frequency of radiation (Hz).

By introducing the concept of the quantum, Planck solved the black body radiation problem, but his solution brought with it a host of other questions about the relationship of matter and energy. Some of these questions remain unanswered to this day. They form the foundation and mysteries of modern nuclear theory.

The Photon. $\quad$ The French scientist *Louis de Broglie* (1875–1960) discovered that the quantum has a dual character. The particle of classical physics also has the nature of a wave. The quantum behaves as a particle that carries energy but has no mass and travels at the speed of light. This particle is called a **photon** [see Figure 1–1(c)]. The German physicist and philosopher *Werner Heisenberg* (1901–1976) assigned an interesting property to the photon. His **uncertainty principle** declares that it is a law of nature that the position and velocity of a photon cannot be simultaneously determined to absolute certainty. It is therefore impossible to predict the path of a single photon. However, the statistical distribution of a great number of photons can very accurately be determined by the theory called **quantum mechanics.**

1–2 THE BEHAVIOR OF LIGHT

As we have now seen, there is no simple answer to the question "What is light?" Light behaves sometimes as a particle and, at other times, as a wave. For our purposes, it is sufficient to know that the wave approach should be applied in the study of radiometry, photometry, and linear optics. The quantum approach best explains light emission and absorption.

$\quad$ The mysterious behavior of light cannot be used to explain away experimental results that disagree with calculations. The principles used in optoelectronics are accurate and exact. Any discrepancy that occurs between theory and measurement is due either to improper measurement or, more likely, an inadequate grasp of the theory involved. The purpose of this book is to explain the theories and demonstrate how to apply them.

The Types of Light Waves

Light may be considered an electromagnetic wave traveling through space at an extremely high rate of speed. Two kinds of wave are possible: **longitudinal** and **transverse** (see Figure 1–2). In longitudinal waves, the oscillatory movement is in the direction of propagation, as in sound waves; in transverse waves, the oscillatory motion is perpendicular to the direction of propagation, as in waves on the surface of water. An **electromagnetic wave** is a transverse wave consisting of electric and magnetic field components that are perpendicular to each other and perpendicular to the direction of propagation (Figure 1–2).

FIGURE 1–2   Longitudinal and transverse waves.

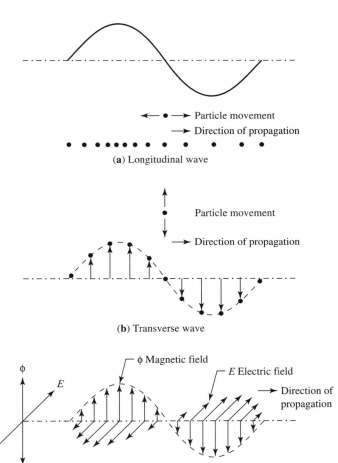

(**a**) Longitudinal wave

(**b**) Transverse wave

(**c**) Electromagnetic transverse wave

To understand waves thoroughly, it is helpful to define the terms that are commonly used in describing all oscillating and wave phenomena. The **period (T)** refers to the time interval required for one oscillation cycle to complete itself (see Figure 1–3). The **frequency (f)** is the number of oscillations that occurs in one second. The inverse relationship between period and frequency is well known:

$$f = 1/T \tag{1-2}$$

where T = period (s)
    f = frequency (Hz).

The third characteristic of oscillation is **amplitude,** which is the maximum absolute value attained by the disturbance of a wave. Amplitude of electromagnetic radiation is a complex combination of magnetic and electric field components. It cannot be determined from direct measurement, as can, say, the swing of a pendulum. In electromagnetic radiation, we deal with the power propagated by the oscillating wave, which is proportional to the square of the amplitude.

EXAMPLE 1–1

Krypton 80 is a gas used to establish the standard of length in the United States. It emits a frequency of 4.948865×10^{14} Hz. Find the period of this oscillation.

Solution
$T = 1/f = 1/4.948865 \times 10^{14} = 0.2020665 \times 10^{-14}$ (s). As you can see, oscillation of light involves incredibly high frequencies and very short periods.

Relationships and Equations

As the oscillation spreads in a medium, it forms a wave such as the one depicted schematically in Figure 1–3. The **wavelength (λ),** which is the distance

FIGURE 1–3   Amplitude and period.

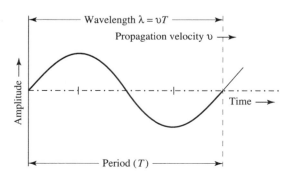

the wave travels during one period, is completed as follows:

$$\lambda = vT = v/f \qquad \textbf{(1–3)}$$

where λ = wavelength (m)
v = velocity of light (m/s).

The velocity of light in a vacuum

$$v = 2.997924574 \times 10^8 \text{ (m/s)}$$

is considered to be one of the fundamental and invariant constants of nature. No longer is the mystic medium "ether" considered necessary to carry electromagnetic radiation. The electromagnetic wave just travels in space.

EXAMPLE 1–2

How many krypton 80 wavelengths are in a meter?

Solution
From Example 1–1, we see that krypton 80 has a period of $0.2020665 \times 10^{-14}$ seconds. This corresponds to a wavelength of

$$\lambda = (0.2020665 \times 10^{-14})(2.9979246 \times 10^8) = 0.6057802 \times 10^{-6} \text{(m)}$$
$$= 605.7802 \text{ (nm)} \quad \text{(an orange-yellow color)}$$

Thus, the number of wavelengths (N) in a meter is

$$N = 1/0.605.7802 \times 10^{-6} = 1.6507638 \times 10^6 \text{ wavelengths}$$

At one time the National Bureau of Standards (now National Institute of Standards and Technology) defined the meter as 1,650,763.73 wavelengths of krypton 80 emission. Now, the meter is defined as the distance that light travels during a time interval of $1/v$ seconds. This measure demonstrates our dependence on and trust in the stability of nature's constants.

The Electromagnetic Spectrum

The **electromagnetic spectrum,** shown in Figure 1–4, ranges from very long radio waves to extremely short cosmic rays. In optoelectronics, we will be concerned with infrared light, visible light, and ultraviolet radiation of wavelengths that range from 1 mm to 1 nm.

Radiometry and Photometry

Optoelectronics is the science of electromagnetic radiation itself and the electromagnetic energy field generated by the radiation sources. **Radiometry,** the study of the properties and characteristics of electromagnetic radiation,

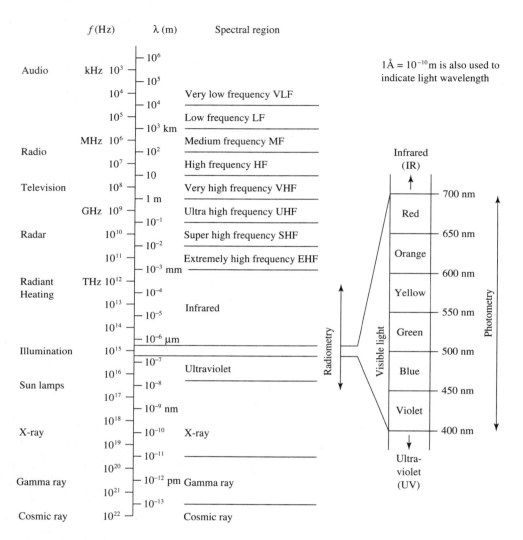

FIGURE 1-4   The electromagnetic spectrum.

describes the radiation field, how the radiation reacts with devices and receivers, and the characteristics of radiation sources and receivers. We will limit our discussion here to the frequencies and wavelengths from infrared to ultraviolet, even though the field of radiometry covers a much wider frequency spectrum. In radiometry, we measure radiation with electronic devices, and the results of such measurements are expressed in commonly employed physical units (watts, watts per square meter, and so on). The science of radiometry is relatively new. It originated about the beginning of the century with the onset of electronics technology.

The spectrum of radiometry includes wavelengths from 400 to 700 nm, the range of visible light. Operating within this range are many optoelectronics devices whose main purpose is to interact and communicate with humans. The predominant receiver in this range is, therefore, the human eye. The science that deals with visible light and its perception by human vision is called **photometry.**

Photometry, in contrast to radiometry, is a much older science, generated by scientists and artists in the nineteenth century. There are fundamental differences between radiometry and photometry. One of the most significant involves the receiving or measuring device. In radiometry, measurements are made with objective electronic instruments; in photometry, the measuring is achieved by the human eye. Another difference is that, in photometry, we use different units to describe the radiation: the light power is measured in lumens, instead of watts, and the power density per area in lux, rather than watts per square meter.

Unfortunately, human vision is a complex and subjective psychophysical process. We literally cannot determine if everyone experiences the color blue in the same manner, nor can we describe red to a color-blind person who has no direct visual knowledge of it. In order to overcome the subjective factor of vision and to establish a well-defined relationship between radiometric and photometric units, the **International Standard Organization (C.I.E.)** has produced the **Standard Observer Curve,** which represents the characteristics of average human vision and defines the relationship between photometric and radiometric units. This curve, with some other characteristics of human vision, is discussed on page 14.

In this book radiometric units are presented with a subscript R; photometric units with a subscript P. The same applies to the equations defining radiometric and photometric relationships. However, please note in cases where an equation applies equally to radiometry and to photometry, the subscripts are omitted.

In some literature the subscript L standing for **luminous** is quite often used for photometric units. Luminous means "pertaining to the light." Because this word serves so generally to describe phenomena involving light, we will be using it throughout this text, and will consider it to be interchangeable with the word **photometric.**

1–3   THE NATURE OF HUMAN VISION

Many consider vision to be our most important sense. It allows us to recognize the dangers we face, to perform our daily tasks, and to enjoy the splendor and beauty of our environment. All such tasks are accomplished by an extraordinarily complex recognition system, of which the human eye is the initial link with the world. In our eyes a representative image is formed using light, with electromagnetic radiation of wavelength from 400 to 700 nm as the propagation medium for imaging.

Why is our vision restricted to such a narrow range? The reason is that the sun's radiation peaks at this wavelength and an abundance of energy in this range is available. The design of a complex vision system for this particular range is thus a lot easier than it would be for any other.

Human vision is an immensely complicated phenomenon. It involves optical, biological, chemical, neurological, and psychological processes. Thus, we characterize it as psycho-physical. The literature on the subject is enormous. It is virtually impossible to do it justice in only a few pages. Here, we will describe some attributes of human vision in the language of electronic engineers, focusing solely on a few properties that can be applied to photometric devices. For those wishing to delve more deeply into the general subject of vision, we recommend reading the first reference from the source list at the end of this chapter.

The Human Eye

The human eye may be likened to a charge coupled device (CCD) camera. A CCD camera is similar to an ordinary one, except for the image-recording medium. The usual camera records on film; a CCD camera records on an array consisting of tightly packed photosensitive cells that generate and hold an electric charge, which is proportional to the image illumination. The CCD array is periodically read and the information sent to an evaluation and storage device. The human eye works in this manner. Figure 1–5 shows its main parts together with some pertinent data.

Image-Forming.   Image-forming in the human eye is accomplished by two lenses: the fixed focal length **cornea** and the **variable focal length lens.** The eye focuses an image by adjusting the focal length of the variable focal length lens. The curvature of this lens may be changed by action of the ciliary muscle and suspension ligament. The eye may therefore be thought of as an automatic-focus camera wherein focusing is accomplished by changes in the focal length of the lens, as opposed to change in the lens image distance, which is what occurs in an ordinary camera. In the eye, located between the cornea and lens, is an automatic diaphragm called the **iris,** which closes or opens depending upon the luminance level of the observed object. The iris regulates the amount of light energy passing through the lens.

The image is formed at the **retina,** which is at the rear surface of the eye. Within the retina are about 130 million photoreceptor cells that convert the image to proportional electrical signals, which are then conveyed by a bundle of optic nerves to the brain. There, electrical signals are analyzed, and our perception of the images is conceived.

Where their concentration is highest, in the **fovea,** the receptors are only a little more than a wavelength apart. Therefore, the resolution power, or **acuity,** of the eye is quite high. In fact, an average human eye can resolve

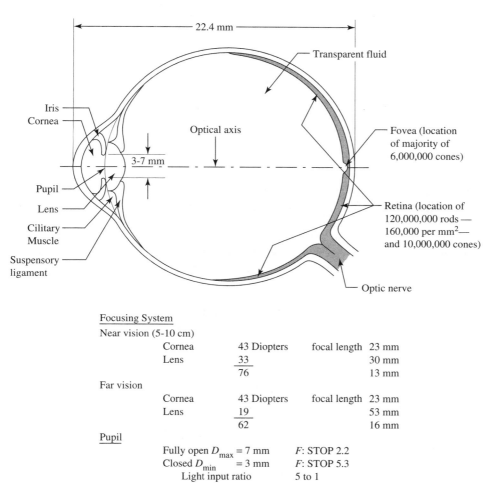

FIGURE 1–5　The human eye.

a line with a width of 0.1 mm at normal reading distance of 250 mm (10 inches). This corresponds to an angle of about 1 minute.

Rods and Cones.　Most characteristics of our vision, such as acuity, color response, wide sensitivity range, and response time, are determined by the intricate structure of the retina. The surface of the retina is covered with two types of photoreceptor cells: called **cones** and **rods.** These two structures differ on three basic levels: sensitivity, color response, and location.

The rods are about 1000 times as sensitive as the cones. They provide vision at very low light levels (night vision). During the day, rods are so saturated with light energy that they do not contribute to image perception. The broad sensitivity range of our vision is due to our high sensitivity and low sensitivity receptors, along with the action of the iris.

The color response of rods is monochromatic and peaks at 505 nm, a blue color. Consequently, at night, all objects look blue and red objects seem almost black. Cones, in contrast, respond to three colors: red, green, and blue. The relative sensitivity curves of the three types of cones is shown in Figure 1–6.

The functioning of the cones explains why modern film and television color techniques utilize red, green, and blue as the primary colors instead of the red, yellow, and blue used by artists. An amazingly sophisticated bio-chemical process determines how the cones respond to color. This process is relatively slow and affects the response time of the eye (see p. 14).

The third difference between the cones and rods concerns their location on the retina. The cones are mostly concentrated in the **fovea,** an area approximately the size of a pinhead located on the optical axis of the eye. The optical axis is an imaginary line through the center of both surfaces of the lens. Our vision achieves its highest resolution and best color response in the fovea. The concentration of the cones is considerably less in areas other than the fovea, which is why our peripheral vision is limited. The rods are spread in much lesser concentration around the retina.

How are the nerve signals sent from the retina to the brain? The rods and cones turn their response to light into electrical signals in an intriguing way. The signals are not of the analog variety with amplitude proportional to the light stimulus but are, rather, a complex pulse-code modulated train. The nerve pulses are at a constant amplitude of about 0.1 V and about 1 ms in duration. The repetition rate changes and is proportional to the logarithm of the stimulus. It is quite remarkable that such a modern communication method was first invented by nature a few million years ago.

Characteristics of Human Vision

A comprehensive description of human vision would probably have to take at least a dozen characteristics into account. Here, we will concentrate on the

FIGURE 1–6   Color sensitivity of cones.

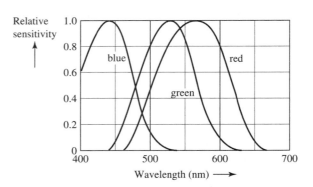

three characteristics that are most important to a designer of optoelectronics devices: the amplitude response, the frequency response, and the response time of human vision. As mentioned, we will speak in electronics vocabulary more so than that of optics on the assumption that the reader will be more familiar with this terminology.

Amplitude Response.   First of all, with respect to the **amplitude response,** it must be said that the human eye is a remarkable instrument. The rods are sensitive enough to respond to a single photon and the eye is capable of operating at radiation density levels up to a few kilowatts per square meter. Putting it in terms of illumination (described on p. 39) the eye can operate at levels from starlight with 0.002 lux up to bright sunlight of 100,000 lux, which encompasses an amplitude range of 1 to 50,000,000. The eye not only covers that huge spread, but is an auto-ranging instrument that adjusts itself to the proper level.

It is easy to understand that, in order to cover such a wide range, a linear response would be impossible. So, the eye has an approximate logarithmic response. The consequence of this fact is that the eye is not a very good instrument for judging levels of absolute illuminance, which is why even highly experienced photographers do not rely on their eyes for the correct exposure, but always use a light meter.

The situation is very different when the eye is used to compare illuminance levels that are similar to each other. Under such conditions, the human eye can detect differences of a few percent. In fact, all earlier photometric measurements were made with instrumentation using the human eye for purpose of comparison, and those measurements were quite reliable and repeatable. This condition is also relevant to optoelectronic displays where similar light sources are next to each other. An intensity difference of about 10% can easily be detected by a human eye.

Frequency Response.   In regard to the second characteristic we have designated **frequency response,** the human vision is also unique. Our eyes not only respond to the amplitude of the received radiation, but at the same time, they also perform frequency analysis of the signal by displaying different wavelengths as different colors. The eye is therefore a spectrum analyzer of electromagnetic radiation over the band of 400 to 700 nm.

The eye's response over this wavelength is not exactly uniform. The cones respond to three colors, as shown in Figure 1–6.

Their combined maximum sensitivity peaks at 555 nm; the rods, at 505 nm (see Figure 1–7).

Therefore, the frequency response of our eye depends upon luminance level, and is usually specified by two curves—one for daylight, or **photopic vision** and another for night, or **scotopic vision.** Those curves (shown in Figure 1–7) describe relative eye sensitivity at different wavelengths, normalized to unity at the most sensitive wavelength. For example, at 500 nm the photopic curve has a value of 0.4, which means that the visual (photometric)

FIGURE 1–7 C.I.E. Standard observer relative sensitivity curves.

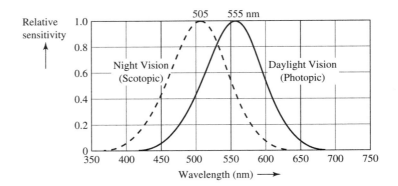

response for the same radiometric stimulus or power is 40% of the response at 555 nm.

The photopic curve is extremely important because it is the key to converting from radiometric to photometric units. For this reason the photopic curve—which has been standardized by International Standards Organization (C.I.E.)—has, as mentioned before, a special name: **C.I.E. Standard Observer Curve.** Because it is often used for calculations, an accurate table and a more usable curve in logarithmic scale is given in Appendix A.

Response Time.   The third characteristic we emphasized is the eye's **response time.** Precisely because the transmittal of optical images to the brain involves chemical processes, the response time of the human eye is relatively slow. Two types of response should interest the optoelectronics designer. One is the capability of the eye to adapt from a high to a low illumination level, a process called **eye adaption.** The other is the eye's ability to respond to rapid change in illumination or, put another way, the response to **flicker.**

The eye's adaption to varying levels of illumination involves a process that is slow and exponential in nature, with a time constant of 2 minutes for cones and 6 minutes for rods. This effect is a somewhat minor factor in the design of optoelectronic equipment. More important is the flicker effect, which is specified by the **critical fusion frequency,** defined as the lowest possible frequency of a 50% duty cycle light signal which the eye will see as a steady light. The eye will notice light variation and will determine flickering of the light at any frequency below that of the critical fusion frequency. Stated in other words, detecting flicker is a measure of the inertia of the eye. The critical fusion frequency for daylight vision is about 40 Hz (or 40 flashes per second); for rods, about 16 Hz.

Flicker plays a highly significant role in many optical devices. For example, our comprehending film or television as a steady moving picture is possible only because of the flicker effect. But how is it that we do not perceive flicker in motion pictures since standard movies are taken at 24 frames a second, which is *below* critical fusion frequency? The answer lies in projection technique: the film is moved at the rate of 24 frames per second, but every frame is projected *twice* before it is moved. Consequently, we see movies at 48 frames per second, which is *above* the flicker frequency.

Because many optoelectronic displays use time multiplexing and strobing techniques, it is essential for optoelectronics designers to understand the flicker effect.

1–4   THE ANGULAR RANGE

Radiometry and photometry deal with the spread of electromagnetic energy in space. Thus, we need a tool to measure that spread. A geometric quantity called **angular range** fulfills this need. Angular range is also referred to as **solid angle** or **space angle.**

On a two-dimensional surface, an angle is most commonly measured in degrees, where one degree is equal to 1/360 of a circle. The number 360 is an arbitrary figure that was established many centuries ago. It is actually a difficult unit to use in mathematical relationships. We will, then, apply a different unit, one that is more logical and suitable for mathematical relationships—the **radian.** A radian is defined as the measure of the angle formed at the center of a circle by two radii cutting an arc whose length is equal to the radius of the circle (see Figure 1–8). The full circle of 360 degrees corresponds therefore to approximately 6.283185 (2π) radians and 1 radian equals approximately 57.29578 degrees.

FIGURE 1–8   Radian and steradian.

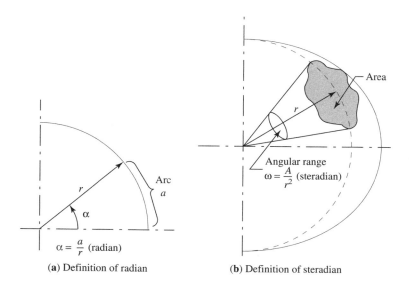

$$\alpha = \frac{a}{r} \text{ (radian)}$$

(a) Definition of radian

Angular range
$$\omega = \frac{A}{r^2} \text{ (steradian)}$$

(b) Definition of steradian

The same principle can be used to measure the spread in space (the angular range). The unit of measure of angular range is the **steradian ω.** The angular range is defined as the ratio of an area formed on the surface of a sphere to the squared radius of the sphere (Figure 1–8), expressed mathematically as

$$\omega = A/r^2 \tag{1-4}$$

where ω = angular range (steradian, sr)
A = area at the sphere surface (m^2)
r = radius of the sphere (m).

The steradian is dimensionless and the maximum spread around a point, or the maximum value of angular range, is

$$\omega_{max} = A_{sphere}/r^2 = 12.5664 \text{ sr}$$

The angular range is commonly used in calculating the coupling between a source and a receiver. A typical example would be to find power emitted from a point source S to a receiver with a receiving area A (see Figure 1–9).

The coupling depends on the angular range from the source S to the area A. In order to determine the angular range, we need to know the corresponding spherical area on the receiving surface. In most cases, however, the **plane** area of the receiving surface is given. Obviously an error occurs when the flat or plane area is used, instead of the spherical surface (see Figure 1–10). For the sake of simplicity, we will calculate the error for a circular cross section cone.

The result (without going into the details of the calculations) is that a circular cone with a half cone angle θ has an angular range

$$\omega = A_s/r^2 = 2\pi(1 - \cos\theta) \tag{1-5}$$

where θ = half-angle of the circular cone (rad or deg).

This angular range of the cone is the correct value. When the plane area A_P is used for calculation, however, a different (and incorrect) value ω' for angular range is determined:

$$\omega' = \pi \tan^2\theta \tag{1-6}$$

which is greater than the correct value. Calculating the maximum value of θ for 1% and 10% error, we find that the simplified method will give less than 1% error when $\theta < 6.59°$ or 0.0415 rad and less than 10% error when

FIGURE 1–9
Source-receiver coupling.

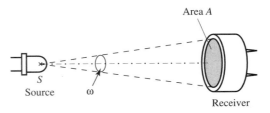

FIGURE 1–10   Angular range calculation from spherical and plane area.

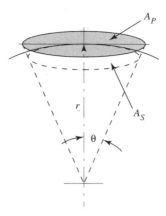

$\theta < 20.15°$ or 0.385 rad. For wider angle cones, the correct equation (1–5) should be used.

EXAMPLE 1–3

Calculate the cone angle for a one-sr circular cone.

Solution
From Equation 1–5, the angular range is

$$\omega = 1 = 2\pi(1 - \cos\theta) \quad \text{or}$$

$$\theta = \arccos(1 - 1/2\pi) = 32.8°$$

Example 1–3 applies to a circular cross-section cone, most common with actual receivers. If the cross section is not circular, the result from the simplified method may have a probable error less than 1% when $\omega \le 0.04$ sr or less than 10% error when $\omega \le 0.4$ sr.

1–5   RADIOMETRIC AND PHOTOMETRIC UNITS AND RELATIONSHIPS

A radiating source emits electromagnetic energy into space, generating an energy field. The disciplines of radiometry and photometry explore and study this energy field by defining units and quantities that describe it, by developing the relationship among these units, and by introducing methods and instrumentation for measuring radiometric and photometric quantities.

Radiometric and Photometric Flux

In physics the term **flux** is often used to describe a flow phenomenon or field condition occurring in space. Figure 1–11(a) presents electric flux – electric

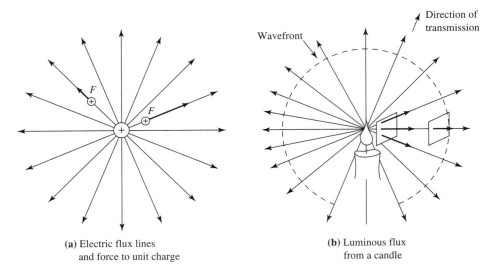

(a) Electric flux lines
and force to unit charge

(b) Luminous flux
from a candle

FIGURE 1–11   Flux in electrical and radiant field.

field lines around a charge depicting the direction and amplitude of a field force applied to a positive unit charge. The direction of the force at any point in space is indicated by the direction of the force line; the magnitude of the force is indicated by the density of the lines.

Figure 1–11(b) presents an example that uses flux lines to show energy transmission from a radiating source. Every line represents a certain amount of emitted power or a number of photons per second. The density of the lines is proportional to the magnitude of the power density emitted in that direction. The arrows on the lines indicate the direction of the spread of radiation. The figure also shows that the power density decreases as the square of the distance from a point source increases.

The symbol for the flux is ϕ; radiometric flux is symbolized by ϕ_R and photometric flux by ϕ_P. In some equations in this text the subscripts ϕ_R and ϕ_P are omitted. As we have mentioned, this emission means that the equation is equally applicable to both radiometric and photometric relationships.

The unit of measure for **_radiometric flux_** is the watt (W); for **_photometric flux,_** the unit of measure is the **lumen (lm).**

The power or flux measured in watts is self-explanatory. The lumen, however, requires further clarification. As pointed out, the photometric units describe the psycho-physical effect of radiation on human vision—they measure how we see light. The eye response (see Figure 1–7, p. 14) depends upon the wavelength we are observing. Therefore, in order to formulate a unique definition of the lumen, we not only have to establish the relationship between radiometric power and photometric response, but we also have to determine the wavelength at which the observation is made. With these factors considered, the ***lumen,*** according to C.I.E. standard, is defined as follows: At 555 nm, at the peak response of photopic vision sensitivity, 1 watt radiometric flux produces 683 lumens of photometric flux, or

$$1 \text{ lumen} = 1/683 \text{ W} \quad \text{at} \quad 555 \text{ nm}$$

We can also define lumen through quantum mechanics; that is, in terms of photon flow rate and **photon energy.** Photon energy (E) at 555 nm is

$$E = \text{h}f = \text{h}v/\lambda = (6.626 \times 10^{-34})(3.00 \times 10^{8}/555 \times 10^{-9}) = 3.582 \times 10^{-19} \text{ J}$$

where h = Planck's constant 6.626×10^{-34} (Js)
    f = frequency of radiation (Hz)
    v = velocity of light (m/s)
    λ = radiation wavelength (m).

One lumen corresponds to

$$1 \text{ (lm)} = 1/683 \text{ (W)} = 1.464 \times 10^{-3} \text{ W} \quad \text{or} \quad \text{J/s}$$

which corresponds to

$$n = 1.464 \times 10^{-3}/3.582 \times 10^{-19} = 4.087 \times 10^{15} \text{ (photons/s)}$$

Therefore, the lumen is equivalent to a flux of 4.087×10^{15} photons per second at the wavelength of 555 nm.

    This definition, together with the C.I.E. Standard Observer Curve, allows us to convert any radiometric flux to a corresponding photometric (luminous) flux using a conversion factor called **efficacy.**

Efficacy and Conversion from Radiometric Flux to Photometric Flux

Efficacy is defined as the ratio of photometric or luminous flux to total radiometric flux from a source:

$$K = \phi_P/\phi_R \tag{1–7}$$

where K = efficacy (lm/W)
    ϕ_P = photometric flux (lm)
    ϕ_R = radiometric flux (W).

EXAMPLE 1–4

A 40-W incandescent bulb typically delivers 460 lumens. What is the efficacy of the bulb?

Solution
To find the efficacy, we have to know the total radiation from the bulb, which typically is not given. However, a pretty good rule of thumb is that, for an

incandescent bulb, about 90% of the power is radiated and 10% is conducted away as heat. Therefore, the total radiometric flux is 36 watts and

$$K = \phi_P/\phi_R = 460/36 = 12.8 \text{ lm/W}$$

The efficacy of the incandescent bulb is low because the filament of the bulb operates at a relatively low temperature and, therefore, most of the radiation occurs in the infrared region.

EXAMPLE 1–5

A typical 40-W fluorescent lamp delivers 3200 lm. Find the efficacy of the fluorescent bulb.

Solution

Once again applying the general rule that about 10% energy is not radiated, we find

$$K = \phi_P/\phi_R = 3200/36 = 88.9 \text{ lm/W}$$

The fluorescent bulb is a much more efficient converter of radiant energy than the incandescent bulb.

Conversion.   We can easily convert monochromatic radiation to photometric flux using the C.E.I. Standard Observer Curve. Efficacy in this case is designated K_λ. It can be determined as follows:

$$K_\lambda = V_\lambda 683 \qquad \textbf{(1–8)}$$

where K_λ = efficacy at wavelength lambda (lm/w)
    V_λ = relative eye sensitivity at wavelength λ
        (from the C.I.E. Standard Observer Curve).

EXAMPLE 1–6

Relative eye sensitivity for green light at 520 nm is $V_\lambda = 0.71$. Therefore the efficacy for this wavelength is

$$K_\lambda = 0.71 \times 683 = 484.9 \text{ (lm/W)}$$

Efficacy of Thermal Radiators.   The most common heated light sources are **thermal radiators.** Also, in physics, they are called **black-body radiators.** Almost every heated metal, including tungsten filament, produces a continuous black-body radiation spectrum. Its spectral radiation pattern depends on radiator temperature. Some black-body radiation patterns are given in Figure 1–12. There we see that low temperature radiators emit more red, and high temperature radiators emit more blue wavelengths. Thus, the radiated light color depends on black-body temperature. For this reason, a term

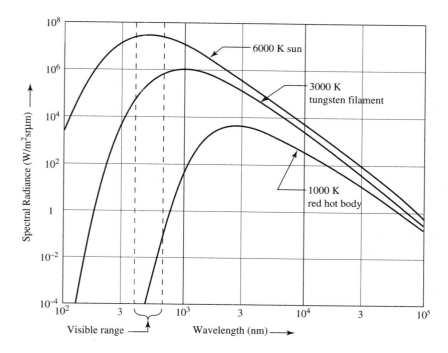

FIGURE 1–12   Blackbody radiation spectrum.

color temperature, which is expressed in degree Kelvin, is used to identify thermal radiators. For given color temperature, the radiation pattern of the source is well defined and the corresponding efficacy can be calculated. The result is given in Figure 1–13. The graph shows that at color temperature of 6000 K, the efficacy reaches a peak because at that temperature the peak of radiation pattern is at the visible range. At other temperatures it is below or above that range.

EXAMPLE 1–7

From Figure 1–12 we can see that a typical tungsten filament has color temperature of 3000 K. This applies to a higher wattage lamp. A 40-watt lamp may have color temperature of 2700 K. What is the efficacy of a tungsten filament lamp?

Solution:
We can read from Figure 1–13 that the efficacy of a black-body radiator at 2700 K is approximately 15 lm/W. Considering several assumptions made about the parameters, this result is in reasonable agreement with the result from Example 1–4.

FIGURE 1-13 Efficacy of blackbody radiator.

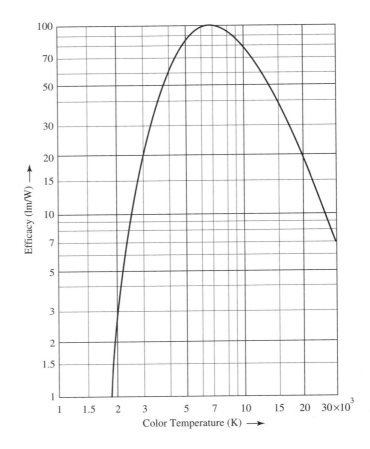

Efficacy of Line Spectrum Radiators.   When the radiation pattern of a source is known, its photometric flux (luminous output) and efficacy can be calculated with the Standard Observer Curve, as demonstrated on p. 23.

A mercury vapor lamp does not emit a continuous spectrum but radiates only on certain discrete wavelengths (see Table 1-1). This type of source is called a **line spectrum radiator.** To find the photometric flux of such source, multiply the radiant flux at each wavelength by the corresponding spectral efficacy:

$$\phi_P = \phi_R K_\lambda \qquad\qquad (1-9)$$

and sum up all radiant fluxes. In Table 1-1 the first (far left) column indicates the wavelength; the second, the radiant flux at that wavelength; the third, the corresponding eye efficiency according to the Standard Observer Curve; the fourth, the corresponding efficacy; and the last (far right), the photometric flux from the radiant flux at that wavelength. Summing the last column will give the total lumen output from the source.

TABLE 1–1
Conversion of Radiant Flux to Luminous Flux in a Mercury Vapor Lamp

Wavelength λ nm	Radiant Flux ϕ_R W	Eye Sensitivity V_λ	Spectral Efficacy $K_\lambda = V_\lambda 683$ lm/W	Luminous Flux $\phi_P = \phi_R K_\lambda$ lm
365	78	0.000	0.00	0.0
408	45	0.001	0.68	31.0
436	70	0.019	13.00	908.0
546	85	0.978	668.00	56,778.0
578	96	0.886	605.10	58,090.0

Total luminous flux   115,807.0 lumens

The data on this table correspond to characteristics of a 1000-watt mercury vapor lamp. Most lamp radiation is outside the visible spectrum. Assuming that the lamp radiates about 90% of input power or 900 W, we can calculate the total efficacy of the source:

$$K = \phi_P/\phi_R = 115{,}807/900 = 128.7 \text{ lm/W}$$

The mercury bulb has higher efficacy than the fluorescent and incandescent light sources. This explains why mercury lights, despite their unpleasant color characteristics, are commonly used for street lights.

Radiometric Energy

Energy is the product of power and time. Thus, **radiometric energy** is

$$N_R = \phi_R t \qquad\qquad \textbf{(1–10)}$$

where N_R = radiometric energy (J, Wh, KWh, etc.)
$\quad t$ = time (s, h, etc.).

Photometric energy is

$$N_P = \phi_P t \qquad\qquad \textbf{(1–11)}$$

where N_P = photometric or luminous energy (lms, lmh, etc.).

A great portion of our electricity bill is used to pay for photometric energy. Efficient generation of this energy may result in considerable savings, as Example 1–8 demonstrates.

EXAMPLE 1–8

Compare the cost of lumen-hour produced by an incandescent and fluorescent source. Assume the cost of electrical energy to be 10 cents/kWh.

Solution

A typical 40-W incandescent bulb costs $0.50, has an average life of 2000 hours, and delivers 420 lumens. The light energy delivered during the lifetime of the bulb is

$$N_P = \phi_P t = 420 \times 2000 = 840,000 \text{ lmh}$$

The cost of producing that photometric energy is

Cost of electricity: $2000 \times 0.04 = 80$ kWh @ $0.10 = $8.00
Cost of bulb: 0.50
Total cost: $8.50

The cost of photometric energy per 1000 lumen-hour is

$$C_P = \$8.50/840 = 0.101 \text{ dollars per 1000 lumen hours}$$

A typical 40-W fluorescent lamp may cost about $5.00, has an average life of 20,000 hours, and delivers 3000 lumens. Photometric energy produced during the lifetime of the lamp is

$$N_P = 20,000 \times 3000 = 60 \times 10^6 \text{ lmh}$$

The cost of producing that energy is

Cost of electricity: $20,000 \times 0.04 = 800$ kWh @ $0.10 = $80.00
Cost of lamp: 5.00
Total cost: $85.00

Cost of photometric energy per 1000 lumen-hours is

$$C_P = \$85.00/60 \times 10^3 = 0.00142 \text{ dollars per 1000 lumen-hours}$$

As seen from this comparison, despite the higher initial cost, the light energy produced by a fluorescent lamp cost about 14%—roughly one-seventh—as much as the energy from an incandescent bulb. This is an important lesson, especially now, when energy conservation is of paramount interest.

Radiometric and Photometric Intensity

Definition of Terms.   Radiometric and photometric intensity describe flux distribution in space. They are defined as follows:

Radiometric intensity (I_R) is the radiometric flux density per steradian, expressed in watt per steradian (W/sr):

$$I_R = \phi_R/\omega \tag{1-12}$$

Photometric or **luminous intensity** (I_P) is the luminous flux density per steradian, expressed in **candelas** (*cd*):

$$I_P = \phi_P/\omega \tag{1-13}$$

One candela equals a luminous flux density of one lumen per steradian. Actually, the correct definition of radiant intensity is $I = d\phi/d\omega$. Therefore, the simplified Equations 1–12 and 1–13 apply only when the flux distribution over the angular range is uniform. This assumption of uniformity is true mostly when the angular range is small.

A short historic note is in order on the candela, an Italian word that literally means a candle. Not too long ago, the luminous intensity was the primary photometric unit from which all other related units were derived. During the nineteenth century, an actual candle was the standard, but, due to lack of repeatability, the standard was then based on the burning of an oil lamp, called a **Heffner lamp,** which, in turn, was replaced by an electric bulb, and more recently, by a glowing, temperature-controlled platinum furnace. The problem with all these standards was that each had a different type of wavelength mix (see p. 21 about color temperature) which made them difficult to compare. Therefore, when advances in electronics made accurate flux measurements possible, the primary photometric standard was redefined from the candela to the flux at 555 nm and the C.I.E. Standard Observer Curve, which explains why the conversion factor—683 lumens per watt—is such an obscure number. This is what makes it compatible to the old candela.

Radiometric (Radiant) Intensity.   The radiant intensity unit of measure is used to characterize a radiation source, specifically the flux distribution from the source. It is appropriate for radiation sources with a relatively small area (*point source*). It is more complex for sources with a wide area, such as diffused fluorescent fixtures.

As the term implies, this unit defines the density of flux emitted by the source. In the visual range, it describes the brightness of the source. A 100 W light bulb and a 100 W automobile headlight deliver the same luminous flux. The headlight, however, is blindingly bright because its flux is concentrated into a much smaller angular range, as illustrated in the following example.

EXAMPLE 1–9

An automobile headlight and a regular clear 100 W bulb both deliver about 1700 lumens. What is the approximate luminous intensity of each source?

Solution

For sake of simplicity, let's assume that the bulb filament is a point source located 55 mm from the base. It radiates evenly in all directions, except through the base. The angular range is therefore

$$\omega = \omega_T - \omega_B = 12.57 - \omega_B$$

where ω_T = total angular range around a point = 12.57 sr

ω_B = angular range to the base.

From Figure 1–14 we can calculate the angular range of the base

$$\omega_B = A_B/d_B^2 = \pi D_B^2/4d_B^2 = \pi \times 25^2/4 \times 55^2 = 0.162 \text{ sr}$$

thus

$$\omega = 12.57 - 0.162 = 12.41 \text{ sr} \quad \text{and}$$

$$I_B = 1700/12.41 = 137.0 \text{ cd}$$

The automobile headlight projects at 24 feet a light pattern as shown in Figure 1–14(b). Let's assume that it evenly illuminates an area that measures 6 feet by 2 feet. From these data we can calculate the angular range

$$\omega_A = A/R^2 = 12.0/24.0^2 = 0.0208 \text{ sr} \quad \text{and}$$

$$I_A = 1700/0.0208 = 81,730 \text{ cd}$$

The luminous intensity of the automobile headlight is considerably higher than the intensity of a light bulb, even when both deliver the same flux.

Our eye definitely notices the difference in luminous intensity, what we think of as source brightness. A source that has twice the luminous intensity as another, however, does not look twice as bright because of the logarithmic response of our eye's perception.

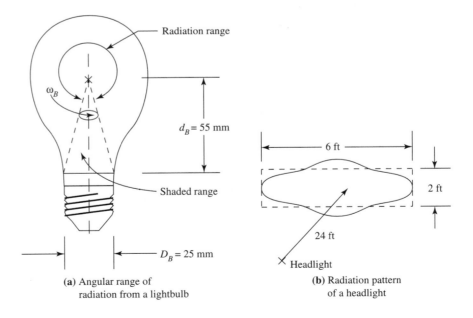

(a) Angular range of
radiation from a lightbulb

(b) Radiation pattern
of a headlight

FIGURE 1–14   Angular range of radiation from a light bulb. Radiation pattern of a headlight.

The typical luminous intensities of some familiar sources are given in Table 1–2.

Graphing Radiant Intensity.   Radiant intensity is a very useful tool for characterizing a source. A measuring of intensity in all directions from the source will accurately describe the radiation pattern from that source. Because most sources have rotational symmetry around a central axis, the measurements can be made and the results plotted in a plane. Two presentations of such graphs are popular—the **polar radiation profile** and the **linear radiation profile**. The term **radiation pattern** is also applied to these graphs. Figure 1–15 presents a flux distribution and both profiles for a common candle.

The profile indicates the radiant intensity in the direction of angle θ measured from the axis of symmetry. The intensity scale may be in candelas, but it is more often expressed in a relative scale, where the maximum intensity is normalized to unity (or 100%), and intensities in all other directions are relative to this. The latter is by far the more practical method, since the relative profile quite often depends upon the fixture or reflector where the source is mounted and does not depend on the absolute intensity values of the source. The polar form is most appropriate for characterizing light fixtures, since it effectively depicts the flux distribution from the source. A few typical examples are shown in Figure 1–16.

The polar form is by far the more descriptive radiation pattern, but the linear form is widely applied because the total source flux is easier to calculate from the linear plot. Deriving the linear profile from a polar form (or vice versa) should be self-evident.

Common Radiant Profiles in Optoelectronics

Three special profiles shown in Figure 1–17 merit further attention because they have numerous applications in optoelectronics.

Besides being common in optoelectronics applications, these sources also have an important advantage: the shape of their radiation profiles can be described in mathematical terms. They can therefore be analyzed mathematically (see below). Having a good mathematical model for the radiation

TABLE 1–2
Luminous Intensities of Common Sources

Source	Luminous Intensity	Source	Luminous Intensity
Regular LED	2 mcd	Automobile headlight	100,000 cd
High intensity LED	120 mcd	Lighthouse	300,000 cd
100-W incand. bulb	150 cd	Flashbulb	1,000,000 cd

FIGURE 1–15   Flux and intensity profiles of a candle.

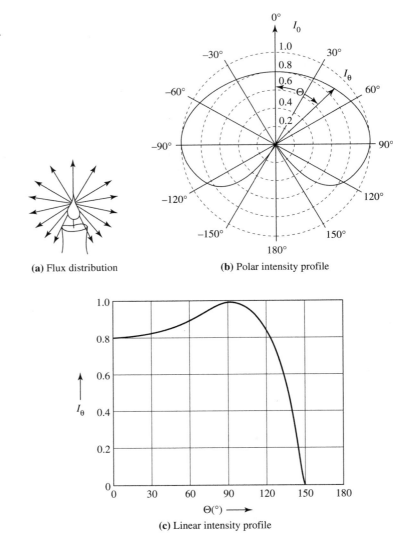

(a) Flux distribution

(b) Polar intensity profile

(c) Linear intensity profile

pattern is important because such a model permits calculation total source flux from a radiation pattern.

Point Source.   The **point source,** shown in Figure 1–17, is the most common source: a small filament in a clear envelope is a typical example. It radiates with equal intensity in all directions:

$$I_\theta = I_0 = \text{constant} \qquad\qquad \textbf{(1–14)}$$

where I_θ = intensity in the direction of angle θ
  I_0 = intensity in the direction of symmetry axes.

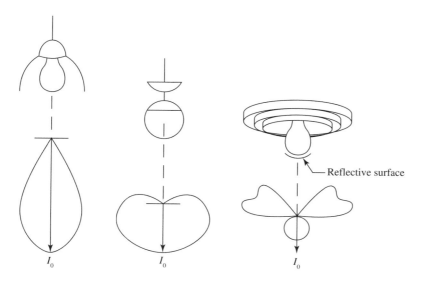

FIGURE 1–16   Intensity profiles of typical light fixtures.

Lambertian Source.   The second source, depicted in Figure 1–17, is called a **Lambertian Source,** and is named after the French scientist Johann Lambert. This type of profile is generated when the light is directed through a diffusing transparent material, or reflected from a rough surface. The light in this case is diffused according to **Lambert's cosine law:**

$$I_\theta = I_0 \cos \theta \qquad \qquad \textbf{(1–15)}$$

The radiation profile according to this equation is a circle. Comparing this profile with that of the point source, we can see that the spread of the radiation is narrower. All radiation is directed only for the angle $\theta = 0$ to 90 degrees. The directionality (or narrowness) of the radiation pattern is a key characteristic of the pattern. To measure the width of the spread, a special term is developed: $\boldsymbol{\theta_{1/2}}$ **angle.** This is the angle θ at which intensity has decreased by a half, or 50%. The smaller the half angle, the narrower and more pointed the radiation pattern.

In the case of the Lambertian pattern, the $\theta_{1/2}$ angle is:

$$I_\theta = I_0/2 = I_0 \cos \theta$$

From here,

$$\theta_{1/2} = \arccos 0.5 = 60° \qquad \qquad \textbf{(1–16)}$$

where $\theta_{1/2}$ = half angle of Lambertian radiation profile.

Exponential Intensity Source.   Figure 1–17(c) illustrates the **exponential Intensity source,** which approximates the radiation pattern from

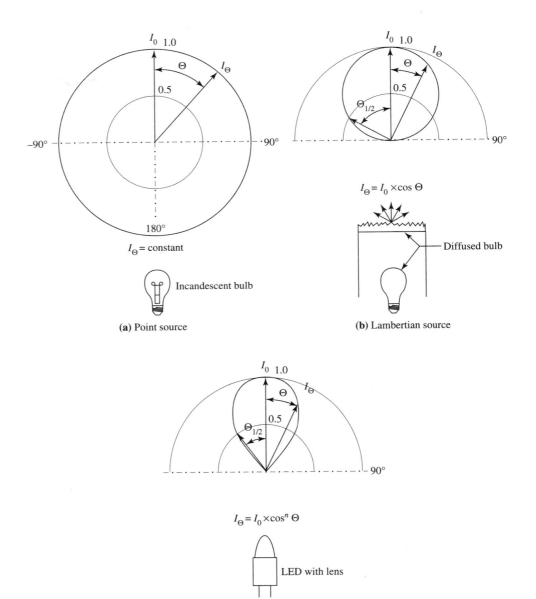

FIGURE 1-17   Intensity profiles of typical optoelectronic sources.

wide- and narrow-beam LEDs. The profile can be specified by the following mathematical relationships:

$$I_\theta = I_0 \cos^n \theta \qquad\qquad (1\text{--}17)$$

where n = radiation pattern exponent.

Here the half angle is

$$\theta_{1/2} = \arccos 0.5^{1/n} \qquad (1\text{–}18)$$

Using the proper value for exponent n, most common LED radiation profiles can be closely approximated using Equation 1–17. Exponent n can easily be calculated when $\theta_{1/2}$ is known:

$$n = |0.301/\log(\cos \theta_{1/2})| \qquad (1\text{–}19)$$

EXAMPLE 1–10

The HP type HLMP-1350 LED has a polar and linear radiation profiles as shown in Figure 1–18. Find the exponent n of the profile, calculate the pattern using Equation 1–17, and compare the actual radiation profile with the calculated profile.

Solution
From the pattern $\theta_{1/2} = 27°$,

$$n = |0.31/\log(\cos 27°)| = 6.00$$

Now the radiation pattern can be calculated and the results shown in polar pattern marked x. It is readily apparent that the calculated pattern using the equation agrees very well with the actual pattern. Therefore, Equation 1–17 gives a good approximation for the LED radiation patterns.

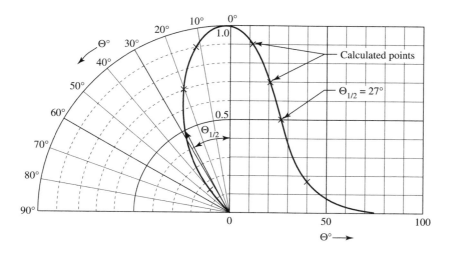

FIGURE 1–18   Polar and linear profile of HP HLMP-1350 LED.

Relationship between Radiant Intensity and Flux

The task of an optoelectronics engineer is to control and manipulate radiant flux. Unfortunately, the measurement of flux, especially over a wider spread, is quite difficult. Measurement of intensity, in contrast, is easy. Therefore, many optoelectronic sources are specified by intensity, rather than by flux, which is of primary interest. To derive flux from a known intensity profile is thus a commonplace task in optoelectronics engineering. As shown by the following, this task is simple for a source with a radiation pattern that exhibits rotational symmetry; that is, the source has a central axis around which the radiation profile stays the same when rotated. Fortunately, most sources are this way.

Determining Flux in a Simple Case. Finding total flux or partial flux for a cone is a matter of simple integration, as Figure 1–19 helps to show us.

The flux into the conical slice d_Θ is

$$d\phi = I_\Theta \, d\omega \quad \text{where}$$

$$d\omega = dA/I_\Theta{}^2 = 2\pi r \, d\theta I_\Theta / I_\Theta{}^2$$

Since $r = I_\Theta \sin \theta$, the flux into the conical slice is

$$d\phi = I_\Theta \, d\omega = I_\Theta 2\pi \, \sin\theta \, d\theta$$

Thus the total flux into an angle θ is

$$\phi_\Theta = \int_0^\Theta I_\Theta 2\pi \, \sin\theta \, d\theta = 2\pi \int_0^\Theta I_\Theta \, \sin\theta \, d\theta \qquad (1\text{–}20)$$

FIGURE 1–19 Flux calculation from the intensity profile.

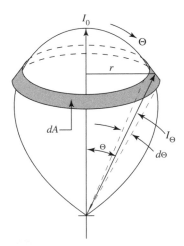

When a mathematical relationship for I_Θ exists, this integral can be solved, as shown in Example 1–11:

EXAMPLE 1–11

Calculate total flux from a Lambertian source that has maximum intensity I_0.

Solution

The intensity function for a Lambertian source is $I_\Theta = I_0 \cos \theta$. Therefore, the integral (1–20) is

$$\phi_\Theta = 2\pi \int_0^\Theta I_0 \cos \theta \sin \theta \, d\theta = 2\pi I_0 \int_0^\Theta \cos \theta \sin \theta \, d\theta$$

$$= 2\pi I_0 |_0^\theta - \cos^2\theta/2 \mid = \pi I_0 |1 - \cos^2\theta| = \pi I_0 \sin^2\theta$$

Since the maximum spread angle in a Lambertian source is 90°, the total flux is

$$\phi_T = \pi I_0$$

Similar calculations have been carried out for point sources and sources with an exponential intensity profile. The results are compiled in Table 1–3. Besides total flux, the flux into a limiting angle θ and half-angle $\theta_{1/2}$ is calculated.

This table contains all information necessary to handle common sources and is a powerful tool for solving most flux calculation problems, as typified in Example 1–12.

EXAMPLE 1–12

The HP LED HLMP-1350 is used in a punched tape reader, as shown in Figure 1–20. The max intensity of the LED is $I_0 = 2$ mcd. Calculate the total flux from the LED and the flux coupled into the photodiode.

TABLE 1–3
Flux Equations for Common Sources

Source	Intensity Profile	Half Angle	Flux into Angle θ	Total Flux
Point	$I_\Theta = I_0$	NA	$2\pi I_0(1 - \cos \theta)$	$4\pi I_0$
Lambertian	$I_\Theta = I_0 \cos \theta$	60°	$\pi I_0 \sin^2\theta$	πI_0
Exponent n	$I_\Theta = I_0 \cos^n\theta$	arccos $0.5^{1/n}$	$\dfrac{2\pi I_0(1 - \cos^{n+1}\theta)}{n + 1}$	$\dfrac{2\pi I_0}{n + 1}$

FIGURE 1–20   Tape
reader configuration.

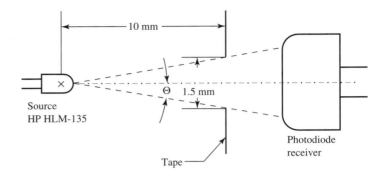

Solution
The LED has an intensity profile $I_\Theta = I_0 \cos^6\theta$ and maximum intensity
$I_0 = 2$ mcd. Using the equation from Table 1–3 for total flux, we obtain

$$\phi_\mathrm{T} = 2\pi I_0/(n + 1) = 2\pi(2 \times 10^{-3})/(6 + 1) = 1.80 \text{ mlm}$$

From the geometry of the punched tape reader, we can calculate the angle of
flux that passes through the punch hole. Thus

$$\theta = \arctan(D/2d) = \arctan(1.5/2 \times 10) = 4.29°$$

Again, using the equation from the Table 1–3, we can find the flux through
the hole, ϕ_D. Thus

$$\phi_\mathrm{D} = 2\pi(1 - \cos^{(n+1)}\theta)I_0/(1 + n) = 2\pi(1 - \cos^7 4.29°)2 \times 10^{-3}/7$$

$$= 0.0349 \text{ mlm}$$

Determining Flux in a Complex Case.   A problem arises when the intensity profile has a shape that cannot be described with a simple mathematical
relationship. In this case, a well-known numerical method may be used to
perform the integration and to find either the total or the partial flux. In nu-
merical integration, the flux into a conical slice $d\theta$: $d\phi = I_\Theta 2\pi \sin\theta \, d\theta$ is replaced by the expression $\Delta\phi = I_\Theta 2\pi \sin\theta \, \Delta\theta$, where $\Delta\phi$ represents the flux
emitted into a slice with the width $\Delta\theta$ at the angle θ. Using the linear intensity profile, divide angle θ into slices of width $\Delta\theta$. I_Θ for each slice can be read
from the profile, and flux into this slice $\Delta\phi$ computed and summed for total
flux. The procedure is described in Example 1–13, with the help of Figure 1–21, which shows total flux from HP HLMP-135 LED with intensity
profile $I_\omega = I_0 \cos^6\omega$ being calculated with numerical integration. The result
is compared with the calculation obtained with the equation from Table 1–3
(see Example 1–12).

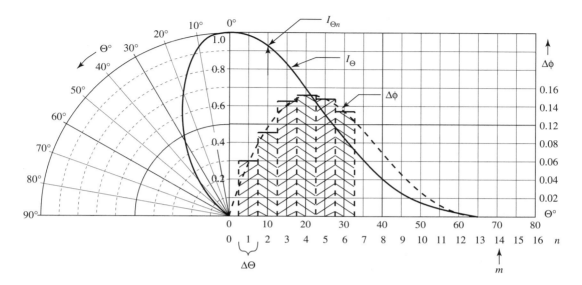

FIGURE 1–21   Computing flux by numerical integration.

EXAMPLE 1–13

For numerical integration, the I_Θ graph has to be divided into m equal slices with width $\Delta\theta$. The higher the number of slices, the more accurate the integration. To achieve accuracy within a few percentage points, at least four slices should be used in angle $\theta_{1/2}$. In this case

$$\theta_{1/2} = 27° \text{ and minimum } \Delta\theta = 27/4 = 6.75°; \text{ selected } \Delta\theta = 5.0°$$

The total number of slices (m) should be selected so that it covers the entire spread of radiation, in this case 70°. Thus, the number of slices (m) is:

$$m = 70/5 = 14$$

For every slice, the spread of flux into this slice is calculated using equation

$$\Delta\phi = I_{\Theta n}\omega_n \qquad \textbf{(1–21)}$$

where $I_{\Theta n}$ = relative intensity at the middle of the slice n
　　　　(read from the linear intensity profile)
　　ω_n = angular range of the slice n (sr).

The angular range of slice n is

$$\omega_n = 2\pi\Delta \sin(n\Delta) \qquad \textbf{(1–22)}$$

where n = slice number
　　Δ = slice width (rad)　　$\Delta(\text{rad}) = \Delta(\text{deg})\pi/180.$

The total flux is the sum of fluxes in all slices plus half of the flux in slice 0

$$\phi = \{0.5I_{\theta 0}2\pi \sin \Delta + \sum_{1}^{n-1} I_{\theta n}2\pi\Delta \sin(n\Delta)\}I_0 \qquad \textbf{(1-23)}$$

where I_0 = absolute intensity at $\theta = 0$ (cd, w/m^2). The computation results are compiled in Table 1–4.

Since $I_0 = 2.0$ mcd, total flux from the LED is $2 \times 0.907 = 1.814$ mlm. In Example 1–12, we calculated the flux using the equation from Table 1–3. The result was 1.80 mlm. As we can see, there is strong agreement between the two methods.

To simplify the calculation, a Basic program for numerical flux integration was developed. It is given in Appendix A.

Optical Transfer Function and Numerical Aperture

The task of an optoelectronics designer is to control the radiant flux—to direct it from a source to a receiver in the most efficient way. In most cases, it is desired to couple as much flux from the source into the receiver as possible. The **Optical Transfer Function** (OTF) is a term that expresses the efficiency of the coupling; the **Numerical Aperture** (NA) is a term used to calculate the OTF. Strictly speaking, neither of these concepts belongs in a discussion on radiometric and photometric units and relationships. They are introduced here because they deal with flux distribution.

TABLE 1–4
LED HLMP-135 Flux Calculation

n	$\theta = n\Delta$	I_θ	ω_n	$\Delta\phi = I_\theta\omega_n$
1	5	0.98	0.048	0.047
2	10	0.91	0.095	0.087
3	15	0.81	0.142	0.115
4	20	0.69	0.187	0.129
5	25	0.54	0.231	0.128
6	30	0.42	0.274	0.116
7	35	0.30	0.314	0.096
8	40	0.20	0.352	0.071
9	45	0.13	0.388	0.048
10	50	0.07	0.420	0.030
11	55	0.04	0.449	0.016
12	60	0.02	0.475	0.001
13	65	0.00	0.497	0.000

$$\Delta\phi/I_0 = 0.883$$
$$\text{Half of slice 0 flux} = 0.024$$
$$\text{Total } \phi/I_0 = 0.907$$

OTF is defined as

$$\text{OTF} = \phi_r/\phi_s \tag{1–24}$$

where ϕ_s = total flux from a source (W or lm)

ϕ_r = flux coupled into the receiver (W or lm).

The OTF, therefore, is a dimensionless quantity, with a range from 0 to 1. Zero means that no flux from the source is coupled into the receiver; one means that all the flux from the source is coupled into the receiver.

Calculating OTF.   In cases where the polar intensity profile of the sender is known, OTF can be readily calculated using equations given earlier in this chapter.

EXAMPLE 1–14

Calculate OTF for the source-receiver relationship shown in Figure 1–22 for a source

(a) with $I_\theta = I_0 \cos^6\theta$ profile and
(b) with Lambertian profile.

Solution (a)

The coupling angle θ from source to receiver is

$$\theta = \arctan D/2d = \arctan 0.1 = 5.71°$$

Using equations from Table 1–3, the flux into a cone of 5.71° is

$$\phi_\theta = I_0 2\pi(1 - \cos^{n+1}\theta)/(n + 1)$$

The total flux from the source is

$$\phi_{\text{TOT}} = I_0 2\pi/(n + 1)$$

The OTF is

$$\text{OTF} = \phi_\theta/\phi_{\text{TOT}} = 1 - \cos^{n+1}\theta = 1 - \cos^7 5.71° = 0.0342$$

Only 3.4% of the source flux is coupled into the receiver.

FIGURE 1–22   Source-receiver relationship.

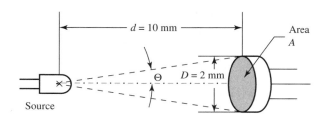

Solution (b)

For source with Lambertian distribution, we have

$$\text{OTF} = \phi_\Theta/\phi_\text{TOT} = I_0\pi\,\sin^2\theta/I_0\pi = \sin^2\theta = \sin^2 5.71° = 0.0099$$

Because the Lambertian source has wider flux distribution, only about 1% of source flux is coupled into the receiver.

For a source with $I_\Theta = I_0\cos^n\theta$ profile, OTF can be easily calculated from the coupling angle θ:

$$\text{OTF} = 1 - \cos^{(n+1)}\theta \tag{1–25}$$

The OTF calculation is one of the most frequently performed in optoelectronics design. To simplify the calculation, the special term **Numerical Aperture** is used.

$$\text{NA} = \sin\theta \tag{1–26}$$

where θ = half of the cone angle to the receiver (deg or rad), as shown in Figure 1–22.

Numerical Aperture is also a dimensionless quantity ranging from 0 to 1. Since, for small angles, the sine and angle, in radians are almost equal, the approximate value for numerical aperture can be calculated by the simplified equation

$$\text{NA} \approx D/2d \tag{1–27}$$

where D = diameter of the circular receiver (m)
d = source receiver distance (m).

When the angle θ is less than 8 degrees, the error in the calculation is below 1%, which represents an accuracy that is quite acceptable in the optoelectronics field. Also, there are cases where numerical aperture has to be calculated from the circular area of a receiver, as,

$$\text{NA} = \sqrt{A/\pi}/d \tag{1–28}$$

where A = area of a circular receiver (m²).

The main application for Numerical Aperture is to simplify the calculation of the optical transfer function. It is especially useful in the case of a source with a Lambertian profile where

$$\text{OTF} = \phi_r/\phi_S = I_0\pi\,\sin^2\theta/I_0\pi = \sin^2\theta = (\text{NA})^2. \tag{1–29}$$

For a point source,

$$\text{OTF} = \phi_r/\phi_S = I_0 2\pi(1 - \cos\theta)/4I_0\pi = (1 - \cos\theta)/2 = \sin^2(\theta/2) \tag{1–30}$$

For a small angle,

$$\sin(\theta/2) \approx (\sin\theta)/2 \quad \text{and therefore}$$

$$\text{OTF} = (\text{NA})^2/4 \tag{1–31}$$

Sources with $I_\Theta = I_0 \cos^n\theta$ profile do not lend themselves to a simplified calculation using numerical aperture.

As a rule, most efficient use of radiant flux is desired in a great majority of optoelectronics applications. Thus, the maximizing of OTF should be every designer's goal. Maximization is accomplished by increasing the NA, determined by the dimensions of receiver area and receiver–source distance. Often, there are physical constraints that limit their variations. Nevertheless, by using a lens between the sender and receiver, the designer can introduce another element into the coupling link by which the OTF can be greatly improved. This situation is discussed in Chapter 2.

Radiant Incidance and Illuminance

Definition of Terms.   Radiant intensity defines flux distribution in space, **radiant incidance** *(E_R)* its distribution on a surface:

$$E_R = \phi_R/A \tag{1–32}$$

where E_R = radiant incidance (W/m^2)
$\quad \phi_R$ = radiant flux (W)
$\quad A$ = area of flux distribution (m^2).

This definition assumes that the flux is evenly distributed over the area.

In the case of luminous flux the quantity is called **illuminance** *(E_P)* and the same definition applies, except that the illuminance has a special unit assigned to it, the **lux** *(lx)*. Thus

$$E_P = \phi_P/A \tag{1–33}$$

where E_P = illuminance (lm/m^2 or lx)
$\quad \phi_P$ = luminous flux (lm).

The lux is an illumination wherein one lumen of luminous flux is evenly distributed over an area of one square meter.

Beside lux, the **foot-candle** *(fc)* is a commonly used unit for illumination, especially among illumination engineers. The foot-candle is an old English unit and is an illumination of one lumen of luminous flux evenly distributed over an area of one square foot. Since the square foot is a smaller area than a square meter, the illumination of one foot-candle is greater than that of one lux:

1 lux = 0.09290 foot-candles
1 foot-candle = 10.764 lux.

Measuring Illumination. Illuminance indicates the flux distribution over an area. It is a term that describes the visibility of objects: a surgeon needs more illumination than what is available in a parking lot. Consequently, illumination engineers are concerned with this concept, since their job is to make objects visible and create visual impressions. Table 1–5 displays typical illuminations for some ordinary places and conditions.

This table not only demonstrates the wide variance of illumination, it also shows the beautiful adaptability of human vision, since we are able to respond at the extremes listed here.

Illumination is one of the simplest photometric quantities to measure. An illumination meter, often called a **light meter,** is required for this measurement. The meter is placed at the surface to be measured, with the probe area pointed at the light source, and the illumination is read from either an analog indicator or a digital display. An illumination meter is easy to design: in the simplest case a photodiode and a microammeter are all that are needed. Figure 1–23 shows a representative illumination meter.

Measuring illumination is the most usual of optometric functions because it is easy to perform and, from it, other pertinent optometric or radiometric quantities can be derived. Here, for example, is a simple relationship of flux falling to an area:

$$\phi = EA \qquad\qquad (1\text{–}34)$$

Relationships among Measurements. Incidance, illumination, and radiant intensity relationships in the case of a point source are shown in Figure 1–24.

Illumination on surface A is

$$E = \phi_S/A$$

where ϕ_S = flux from the source (lm)
$\quad A$ = surface area (m^2).

TABLE 1–5
Typical Illumination Levels

Condition	Illumination (lx)
Sunlight	100,000
Dull day	1000
Merchandise display	1000
Reading area	500
Parking area	50
Moonlight	0.4
Starlight	0.002

FIGURE 1–23
A versatile illuminance,
radiance, and luminance
meter. (Courtesy
Tektronix, Inc.)

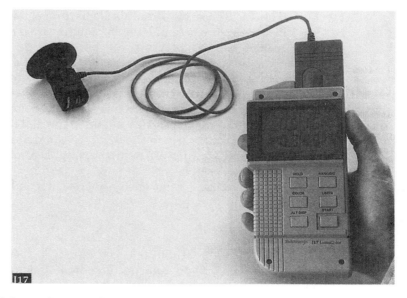

The flux can be calculated from the angular range ω where

$$\omega = A'/d^2 = A \cos \alpha/d^2$$

thus

$$E = I \cos \alpha/d^2 \qquad \qquad \textbf{(1–35)}$$

where d = source to surface distance (m)
    α = angle between the light ray and surface normal.
    I = source intensity (w/sr or cd)

In the case in which the flux from the source is perpendicular to the surface,
$\cos \alpha = 1$ and

$$E = I/d^2 \qquad \qquad \textbf{(1–36)}$$

where E = incidance or illumination (w/m^2 or lx).

FIGURE 1–24 Illumination
from a point source.

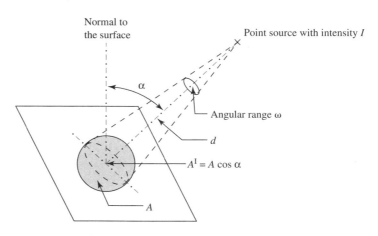

The incidance or illumination is proportional to the source intensity and inversely proportional to the square of source-surface distance. Equation 1–36 is especially significant because it gives a simple relationship between the intensity and incidance or illumination. It allows us to derive the value of intensity, which is a difficult quantity to measure, from simple measurement of incidance or illumination and distance. The equation applies to conditions where the source area is small, as compared to the distance d: that is, where the source looks like a point source. The fact that the illumination decreases proportionally to the distance squared is quite significant. It points to the principle that for efficient coupling, the source and the receiver should be as close as possible.

A use of the relationships presented above is illustrated in Example 1–15.

EXAMPLE 1–15

A miniature bulb, GE 86 (6.3 V, 0.2 A, $I_B = 0.4$ cd), is 20 mm away from a Motorola MRD 500 photodiode. The diode has a responsivity of 5.0 μA/mW/ cm^2 for a tungsten source. Find the photodiode output.

Solution

At the photodiode surface, the illumination is

$$E_D = I_B/d^2 = 0.4/(20 \times 10^{-3})^2 = 1000 \text{ lx} \quad \text{or} \quad \text{lm/m}^2$$

The photodiode responsivity is given in mW/cm^2. Thus, the luminous flux has to be converted to radiant flux. The efficacy of black-body radiators is given in Figure 1–13, p. 22. Assuming a color temperature about 2200 K, the efficacy from this curve is 4 lm/W. Therefore, the illumination of 1000 lm/m^2 corresponds to radiant incidance

$$I_R = 1000/4.00 = 250 \text{ W/m}^2 = 250/100^2$$
$$= 25.0 \times 10^{-3} \text{ w/cm}^2 \quad \text{or} \quad 25.0 \text{ mW/cm}^2$$

From here, photodiode output is

$$25.0 \text{ mW/cm}^2 \times 0.5 \ \mu\text{A/mW/cm}^2 = 12.5 \ \mu\text{A}$$

Radiant Sterance and Luminance

The most important aspect of radiation, especially radiation in the visible range, is the fact that it allows us to see objects. Actually, what we are seeing is not the luminous flux from the source, not the illumination or intensity of flux on the surface of an object, but the flux that is reflected or radiated from a source and received by our eye. This reflection or radiation allows us to identify the objects by their brightness, color, and surface structure. This reflection or radiation is called **radiant sterance** (L_R) or, in the case of visible light, the **luminance** (L_P). In fact, most optical devices that use a lens, including the human eye, respond to luminance or sterance. When we photograph a scene with a camera loaded with black and white film, we record the luminance of the objects in the scene. Often this term is used interchangeably with the term **brightness**. The terms are quite similar, the difference being that the brightness refers to our eye's and brain's psycho-physical response to luminance, which happens to be a logarithmic, instead of a linear function.

Measuring Radiant Sterance and Luminance.   The units of measure for radiant sterance and luminance are as follows:

For radiant sterance, (L_R): W/sr/m²

For luminance, (L_P): cd/m² or lm/sr/m²

The unit has some similarity to intensity, which is measured in w/sr or lm/sr and which is described also as the "brightness" of a point source. The term luminance or radiant sterance carries this idea to a wide emitting or reflecting surface, or to an **extended source,** as it is often called. Thus, they are defined as "brightness per area."

Luminance of a Surface.   The question is how "bright" or luminous an illuminated surface A looks to an observer at O. First, let us look at a case where the surface is an active emitting source (seen in Figure 1–25).

The luminance or sterance is defined as follows:

$$L = I_\Theta/A' = I_\Theta/A \cos \theta \qquad\qquad \textbf{(1–37)}$$

FIGURE 1–25   Luminance of a surface with Lambertian pattern.

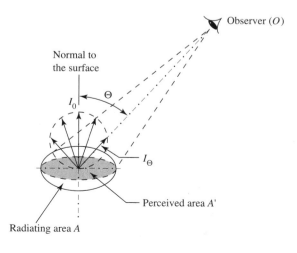

Observer (O)

Normal to the surface

I_0

Θ

I_Θ

Perceived area A'

Radiating area A

where L = luminance (cd/m^2 or lm/sr/m^2) or sterance (W/sr/m^2) of the surface A

A = radiating area (m^2)

A' = area perceived by the observer O (m^2)

I_θ = source intensity in the direction of the observer (cd or W/m^2).

Lambertian Source.   In a case in which the source radiation has a Lambertian profile, $I_\Theta = I_0 \cos \theta$, Equation 1–37 can be written

$$L = I_0 \cos \theta / A \cos \theta = I_0 / A \qquad (1\text{–}38)$$

Because the intensity of a Lambertian source and also the perceived surface area decrease proportional to $\cos \theta$, their ratio or the sterance or luminance is the same in all directions. This equation explains why sun surface, or the surface of a frosted light bulb or fluorescent light tube has the same brightness over the entire surface: the light distribution profile from each source is Lambertian. This is usual for most radiators. Table 1–6 gives luminance values for some common objects.

EXAMPLE 1–16

A 60-W frosted bulb, which has a bulb diameter of 60 mm and radiates through an angular range of 11.8 sr (see Figure 1–14, p. 26), delivers 850 lumens. Find the luminance.

Solution

The luminous intensity of the bulb is:

$$I_\mathrm{P} = \phi_\mathrm{P}/\omega = 850/11.8 = 72.0 \text{ (cd)}$$

The bulb radiates through an area that is proportional to radiated angular range over total angular range of a sphere:

$$A = A_{\mathrm{SP}}\omega/4\pi = \pi D^2 \omega/4\pi = 0.06^2 \times 11.8/4 = 0.0106 \text{ (m}^2)$$

Thus the luminance of the bulb is

$$L_\mathrm{P} = I_\mathrm{P}/A = 72.0 \ / \ 0.0106 = 6{,}780 \text{ (cd/m}^2)$$

TABLE 1–6
Luminance Values of Common Objects

Source	Luminance (cd/m^2)
Sun	1.6×10^9
Clear incandescent filament (500 W)	11×10^6
Incandescent lamp, frosted (500 W)	300×10^3
Fluorescent lamp	10×10^3
Candle flame	5×10^3

The value is considerably less than luminance given in Table 1–6 because it is calculated for a 60 W bulb as compared to a 500 W bulb. The latter is definitely brighter.

The term luminance applies also to passive surfaces that are illuminated, often called **extended sources.** The luminance of such a source or surface is caused by the reflection of the incident flux. The reflecting properties of the surface also determine the color of the surface. In a blue surface, all wavelengths except the blue are absorbed and only wavelengths representing the blue are reflected back.

The luminance of an extended source therefore depends upon two factors: the illumination and the **reflection coefficient** of the surface. The relationship between illumination and luminance is easy to derive for surfaces with a Lambertian reflection pattern. The reflection coefficient can be defined as

$$R = \phi_{REF}/\phi_{INC} \tag{1–39}$$

where R = reflection coefficient
  ϕ_{REF} = reflected flux (W or lm)
  ϕ_{INC} = incident flux (W or lm).

The incident flux can be calculated from the illumination of the surface

$$\phi_{INC} = AE$$

where A = surface area (m^2)
  E = illumination (W/m^2 or lm/m^2).

Thus the reflected flux is

$$R_{REF} = RAE$$

Using the equation in Table 1.3, we can calculate the intensity from the total flux:

$$I_{REF} = \phi_{REF}/\pi = RAE/\pi$$

and using Equation (1–39), we can calculate the luminance perpendicular to the surface:

$$L = I_{REF}/A = RE/\pi \tag{1–40}$$

When the reflection pattern is Lambertian, the luminance in all directions is the same. Since most surfaces are Lambertian reflectors, we can easily confirm this law by observing actual objects. A typical example is the moon's surface, which has uniform luminance. The same equation also explains why the luminance or brightness of objects does not change with distance or direction of the observer.

Photography Application.   As mentioned before, luminance and radiant sterance are the measured or recorded quantities in systems that use a lens. A photographic camera and the human eye are the most common representatives of this group. A photodiode, facing a source, responds to illumination or radiant incidence. This is an important difference that has to be considered when different sensors are used.

As photographers know, object illuminance can be measured two ways, using either a reflected or incident light meter. The reflected light meter is used at the camera and pointed toward the object. It directly measures object illuminance. The incident light meter is used at the object and it actually measures illumination and computes luminance using Equation 1–40. The reflection coefficient of the average photographic object is 18% and thus, for camera exposure purposes, the luminance is

$$L_{\text{photo}} = 0.18 \times E/\pi = 0.0573 \times E \tag{1–41}$$

The exposure is inversely proportional to the luminance. There are, however, two other factors that contribute to the exposure: lens aperture and film sensitivity. Their contribution is expressed in the **light meter calibration rule:**

When the lens aperture (F-NUM) is set to a value that equals the square root of the ISO exposure index, the exposure time is the reciprocal of the object's illuminance expressed in cd/square foot.

EXAMPLE 1.17

A picture is taken on a dull day with illumination of 1000 lux using film with ISO sensitivity of 100. Find exposure time and F-NUM setting.

Solution:
$$\text{F-NUM} = \sqrt{100} = 10$$

The luminance of an 18% reflecting object is

$$L_{\text{PH}} = 0.18 \times 1000/\pi = 57.3 \text{ cd/m}^2$$

Converted to foot candles

$$L_{\text{PH}}' = 57.3/10.76 = 5.32 \text{ cd/ft}^2$$

Thus the exposer time is $t = 1/L_{\text{PH}}' = 1/5.32 = 0.188$ s.

Radiant and Luminous Exitance

Exitance *(M)* describes the flux density from a radiating area. It has the same dimension, W/m² or lm/m², as the radiant incidence or illumination. The explicit difference is that exitance applies to an active (radiating) area, incidence to a passive (receiving) area. Thus,

$$M_{\text{R}} = \phi_{\text{R}}/A \quad \text{and} \quad M_{\text{P}} = \phi_{\text{P}}/A \tag{1–42}$$

where M_R = radiant exitance (W/m^2)
M_P = photometric exitance (lm/m^2)
A = emitting area (m^2).

Relationships Derived from Exitance. From exitance other characteristics may be derived; for example, flux from a wide-area Lambertian source is

$$\phi = MA$$

And for any Lambertian source

$$I_0 = \phi/\pi = MA/\pi \tag{1–43}$$

Figure 1–26 shows a wide-area Lambertian source illuminating a surface. In this case, the following relationships can be developed:
The intensity in the direction of the receiver is

$$I_\Theta = I_0 \cos \theta$$

The flux into the receiver is

$$\phi_r = I_\Theta \omega \quad \text{where } \omega = A_R'/d^2 = A_R \cos \alpha/d^2$$

Thus

$$\phi_r = I_0 \cos \theta A_R \cos \alpha/d^2 = \phi_S A_R \cos \theta \cos \alpha/\pi d^2 \tag{1–44}$$

where ϕ_r = flux into the receiver (W or lm)
ϕ_S = source flux (W or lm)
I_0 = source intensity perpendicular to source surface (W/sr or lm/sr)
θ = receiver direction from the perpendicular to the source (deg or rad)
A_R = receiver area (m^2)
α = receiver angle from the perpendicular (deg or rad)
d = source receiver distance (m).

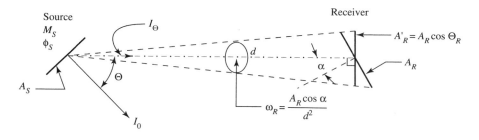

FIGURE 1–26 Illumination from a wide-area Lambertian source.

Equation (1–44) is applicable when source-receiver distance is much greater than source or receiver dimensions. Since illumination on the receiver surface is $E_r = \phi_r/A_R$, we can write

$$E_r = I_0 \cos \theta \cos \alpha/d^2 = \phi_S \cos \theta \cos \alpha/\pi d^2 \qquad \textbf{(1–45)}$$

If receiver and source areas are both perpendicular to the connecting line: $\cos \theta = 1$ and $\cos \alpha = 1$

$$\text{and} \quad \phi_r = I_0 A_R/d^2 \qquad \textbf{(1–46)}$$

$$\text{and} \quad E_r = I_0/d^2 = \phi_S/\pi d^2 \qquad \textbf{(1–47)}$$

Often wide-area sources are defined by their luminance or sterance, which can be expressed $L = I_0/A_S$ or $I_0 = LA_S$. In this Equations 1–44 and 1–45 will take a new form:

$$\phi_r = LA_S A_R \cos \theta \cos \alpha/d^2 \qquad \textbf{(1–48)}$$

$$E_r = LA_S \cos \theta \cos \alpha/d^2 \qquad \textbf{(1–49)}$$

where L = source luminance or sterance (cd/m² or W/srm²).

When source and receiver surfaces are parallel and perpendicular to the flux axes $\cos \theta = \cos \alpha = 1$ and the above equation simplifies to

$$E_r = LA_S/d^2 = L\omega_S \qquad \textbf{(1–50)}$$

where ω_S = angular range of the source subtended by the receiver (sr).

EXAMPLE 1–18

A 100-W incandescent bulb that delivers 1700 lm is mounted into a ceiling fixture that directs 50% of bulb flux downward through a diffusing cover. A desk is 5.5 feet below and 3 feet in front of the fixture. What is the illumination on the desk?

Solution

The geometry of the arrangement is shown in Figure 1–27. The downward flux from the bulb is ϕ_S = 50% of 1700 lm = 850 lm. Since it is a diffusing or Lambertian fixture, the maximum intensity $I_0 = \phi_S/\pi = 271$ lm/sr. From the geometry of the arrangement, $\theta = \alpha = \arctan(x/y) = 28.6°$, and $\cos \theta = \cos \alpha = 0.878$.

The distance between source and receiver is $\sqrt{x^2 + y^2} = 1.91$ m. Using Equation 1–45 we find

$$E_R = I_0 \cos \theta \cos \alpha/d^2 = 271 \times 0.8782^2/1.91^2 = 57.2 \text{ lx}$$

FIGURE 1-27
Example 1-18: fixture and
desk arrangement.

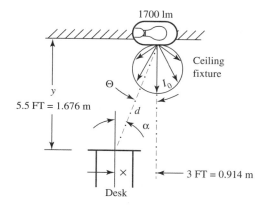

SUMMARY

This chapter describes the basic tool of optoelectronics—radiometry and photometry. Here the nature and short history of radiation phenomena are introduced. Further, the difference between radiometry and photometry is explained. The connecting link between these sciences is that of human vision. Thus, the basic characteristics and properties of the human eye and human vision are introduced. Further, a set of radiometric and photometric quantities and terms are introduced and defined. The thrust of the chapter is in developing the relationships among these quantities, which are the primary tools of optoelectronic design.

RECOMMENDED READING

Williamson, S. J. and Cumming, H. Z. 1983. *Light and Color in Nature and Art.* New York: John Wiley and Sons.

Stimson, A. 1974. *Photometry and Radiometry for Engineers.* New York: Wiley-Interscience Publication.

Härtel, V. 1978. *Optoelectronics: Theory and Practice.* New York: Texas Instruments Electronics Series, McGraw-Hill Book Company.

Hewlett-Packard 1981. *Optoelectronics Fiber-Optics Applications Manual.* New York: McGraw-Hill Book Company.

Sir Isaac Newton. 1979. *Opticks.* New York: Dover Publication, Inc.

PROBLEMS

1. A high efficiency red LED lamp has peak radiation at 635 nm and 3 dB bandwidth at 615 and 653 nm. What is the peak frequency and 3 dB bandwidth?

2. A flashlight projects a 6-inch diameter circle to a wall 4 feet away. Compute the angular range of the beam using plane area cone and exact equation. Find the error when plane area is used.

3. A YAG laser operating at 633 nm has output power of 10 mW. What is the luminous flux from the laser?

4. A 60-watt incandescent bulb delivers 920 lumens at color temperature of 2850 K. What is the radiated flux, and how much power is lost through heat conduction?

5. A 5-watt flashlight with $\phi = 65$ lumens projects a 10-ft evenly illuminated circle at 100 ft distance. What is the luminous intensity of the flashlight?

6. An ultra-bright red LED has center luminous intensity of 100 mcd. The radiant pattern follows the $\cos^n \theta$ law. The half angle is $\theta_{1/2} = 24$ degrees. What is the intensity at 15 degrees?

7. A 2 mm diameter receiver is 10 mm from the source. The source has $I_0 = 200$ mcd. Calculate the flux received by the receiver and OTF when
   a. the source has a Lambertian intensity profile
   b. the source intensity profile is $I_\theta = I_0 \cos^8 \theta$.

8. A light source with $I_0 = 100$ cd is 2 m from a wall. The central axis of the source is perpendicular to the wall. What is the illumination at a point that is 1 m from the central axis when
   a. the source is a point source?
   b. source has a Lambertian intensity profile?

9. The flame of a candle has an area of 2 cm^2 and intensity of 1 cd. What is the luminance of the flame?

10. A 50% reflecting Lambertian surface is illuminated by a source with intensity of 500 cd. The source is 10 m away and located 20 degrees from the perpendicular of the surface. Find the surface luminance.

The answers are in Appendix C.

SURVEY OF PHOTOMETRIC UNITS

During the development of photometry and radiometry many other units, besides the *SI units* (The International System of Units) appearing in this book, have been defined and used. In modern usage, the SI units are preferred, but the others still appear in literature and are used in some applications. A list of their definitions and conversion factors follows.

Wavelength may be measured in angstrom (Å), where

$$1 \text{ Å} = 10^{-8} \text{ m} = 10 \text{ nm}.$$

Luminous flux density is also measured in phots (ph) and foot candles (fc), where

$$1 \text{ ph} = 1 \text{ lm/cm}^2 = 10^4 \text{ lx} = 929 \text{ fc}$$
$$1 \text{ fc} = 1 \text{ lm/ft}^2 = 10.76 \text{ lx} = 0.00108 \text{ ph.}$$

Luminance is also measured in stilbs (sb) and candela per square foot (cd/ft^2) and nit (nt), where

$$1 \text{ nt} = 1 \text{ cd/m}^2 = 10^{-4} \text{ sb} = 0.0929 \text{ cd/ft}^2$$
$$1 \text{ sb} = 1 \text{ cd/cm}^2 = 10^4 \text{ nt} = 929 \text{ cd/ft}^2$$
$$1 \text{ cd/ft}^2 = 10.76 \text{ nt} = 0.0001076 \text{ sb.}$$

Luminance of a Lambertian source is also measured in apostilb (asb), lambert (L), and footlambert (fL) where

$$1 \text{ asb} = 1 \text{ cd/}\pi\text{m}^2 = 0.318 \text{ cd/m}^2 = 10^{-4} \text{ L}$$
$$1 \text{ L} = 1 \text{ cd/}\pi\text{cm}^2 = 3180 \text{ cd/m}^2 = 10^4 \text{ asb} = 929 \text{ fL}$$
$$1 \text{ fL} = 1 \text{ cd/}\pi\text{ft}^2 = 3.42 \text{ cd/m}^2 = 10.76 \text{ asb} = 0.0001076 \text{ L.}$$

Apostilb and lambert make the conversion from exitance to luminance simple. For a Lambertian source with exitance M lm/m^2 the luminance is $L = M/\pi$ when measured in cd/m^2. Using apostilb, the relationship is $L = M$.

2
Elements of
Geometric Optics

Geometric optics is the process of tracing light rays through lenses and mirrors to determine the location and size of the image from a given object. In simpler terms, geometric optics is the science of mirrors and lenses.

Why are opotoelectronics designers interested in geometric optics? Because the task of the optoelectronics designer is to manipulate radiant or luminous flux to achieve maximum optical transfer function, concentrate flux on a given area, and form images from an object. The most effective tools for these purposes are mirrors and lenses, but especially lenses. Thus this chapter covers elementary principles for using single element lenses. (For more complicated systems, students should consult any of the variety of good textbooks on lens optics that are readily available.)

2-1 REFLECTION AND MIRRORS

Mirrors and reflection are the most common geometric optics phenomena encountered in our everyday lives. When light rays from an object strike a polished metallic surface and reflect back, forming an image of the object behind the surface, we have a mirror.

The law of reflection states:

1. The incident ray ϕ_i, the reflected ray ϕ_r, and the line OP, perpendicular to the mirror surface at point O, are all on the same plane.
2. The incident angle θ_i equals the reflected angle θ_r.

From these simple principles, the formation of an image of an object in the front of a mirror is demonstrated in Figure 2–2.

FIGURE 2–1   Law of reflection.

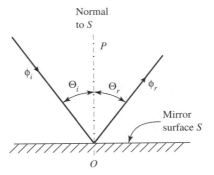

FIGURE 2–2   Forming an image in a mirror.

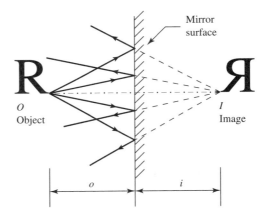

Applying simple geometry and the rule of reflection proves that reflected rays originate from a single point I, which is the image point of the object O. Also, the object distance from the mirror surface o equals the image distance i. There are, however, two interesting properties connected with the image. First, the image is a **virtual** image; it appears at a point where light rays actually never exist. Therefore, we cannot use a plane mirror alone to form a real image from which a permanent replica of an object on a photographic film can be made. For this purpose, we need some other optical system, one that forms a real image. Second, the mirror image is **perverted;** that is, the right and left sides of the image are reversed. For example, the letter R in the mirror appears as Я. At the same time the up and down directions retain their normal positions. Technically, the image is described as being **erect** and **perverted.** This phenomenon is easily explained since the mirror "sees" the object from the side opposite that of the observer. The perversion is important to keep in mind when a mirror is used in an image forming optical system.

The reflected flux ϕ_r is always less than the incident flux ϕ_i by the **reflection coefficient** (R):

$$\phi_r = \phi_i R \qquad\qquad \textbf{(2–1)}$$

where ϕ_r = reflected flux (W or lm)
ϕ_i = incident flux (W or lm)
R = reflection coefficient.

Figure 2–3 shows the value of the reflection coefficient of various metals used for reflectors.

The graph in Figure 2–3 shows that all metals are very good reflectors at infrared wavelengths. Silver and gold are especially excellent with reflection coefficients of over 0.99. The coefficient at wavelengths in the visible range or shorter changes quite rapidly and, therefore, the reflectors for ultraviolet should be selected carefully.

When the surface of the reflecting material is not smooth and polished, the reflection of incident rays is diffused. The degree of diffusion depends on the roughness of the surface, and reflected intensity patterns may form, as shown in Figure 2–4.

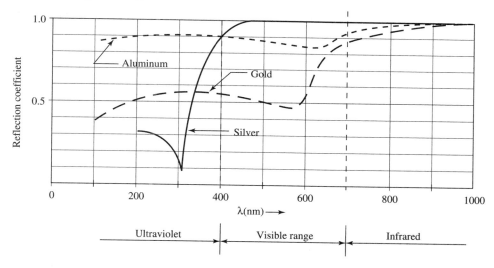

FIGURE 2–3   Reflection coefficient of typical metals.

FIGURE 2–4   Reflection from nonpolished surfaces.

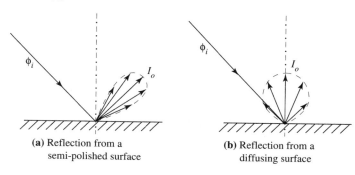

(a) Reflection from a
semi-polished surface

(b) Reflection from a
diffusing surface

The maximum reflected intensity may be calculated when the reflecting pattern of the surface is known. In the case of a randomly rough surface, the pattern is called a **Lambertian pattern** and the maximum intensity I_0 is

$$I_0 = \phi_i R / \pi \tag{2-2}$$

EXAMPLE 2-1

A gold heat shield of 0.5 cm^2 area is in an intense ultraviolet radiation field of 100 nm wavelength at radiant incidence of 0.2 mW/mm^2. How much radiation power is absorbed by the shield?

Solution

The absorbed power is the difference between the incident power and radiated power. The incident power is

$$\phi_i = E_i A = 0.2 \times 0.5 \times 10^2 = 10 \text{ mW}$$

The reflection coefficient for gold at 100 nm is approximately 0.4. Therefore, the surface will radiate

$$\phi_r = \phi_i R = 10 \times 0.4 = 4 \text{ mW}$$

The absorbed power is the difference

$$\phi_a = \phi_i - \phi_r = 10 - 4 = 6 \text{ mW}$$

2-2 REFRACTION AND LENSES

The speed of light in a vacuum — 2.99792457×10^8 m/s — is one of nature's constants. When light travels in another media its speed decreases, depending on the media and wavelength. Christiaan Huygens, the originator of the wave theory of light, showed that at the interface of two media with different speeds of light a **refraction,** or change in the direction of the light ray, takes place. Actually, a much more complex event happens at the interface. In addition to the refraction, part of the flux is always reflected and the wave polarization changes. (Since wave polarization has little significance for optoelectronics application, it is not discussed in this text.)

The amount of refraction and reflection at the material interface depends on the angle of the incident ray and the relative change in the speed of the wave in the media. In practical application the actual speed of light is not employed. Instead, the term **refractive index** is used. The refractive index is the ratio of the speed of light in vacuum to the speed of light in a given media:

$$n = v_v / v_m \tag{2-3}$$

where v_v = speed of light in the vacuum (m/s)

v_m = speed of light in the media (m/s)

n = refractive index.

Table 2–1 gives typical values for some optical materials at 550 nm.

Generally, the denser a material, the higher its refractive index. Also, the value of the index depends on the wavelength, increasing at shorter wavelengths and decreasing at longer ones. A typical example is shown in Figure 2–5.

The change of the refractive index of water at different wavelengths causes the dispersion of the white light into colors, creating a rainbow.

Using the refractive index, the laws for refraction and reflection at the interface of two media can be formulated, as shown in Figure 2–6.

TABLE 2–1
Refractive Index Values for Common Materials

Media	Refractive Index
Vacuum	1.00000
Air	1.00003
Water	1.33
Fused quartz	1.46
Crown glass	1.52
Flint glass	1.62
Diamond	2.42

FIGURE 2–5 Wavelength dependence of refractive index.

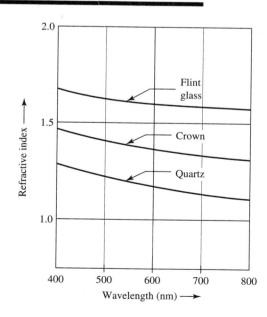

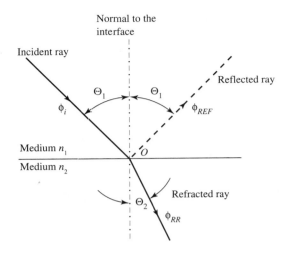

FIGURE 2-6 Reflection and refraction at two media interface.

1. The incident ray, the refracted ray, the reflected ray, and that perpendicular to the interface at point O are all on the same plane.
2. The reflected angle is equal to the incident angle.
3. The refracted angle θ_2 can be calculated from the incident angle using **Snell's law,** expressed as

$$n_1 \sin \theta_1 = n_2 \sin \theta_2 \qquad (2\text{-}4)$$

where n_1 = refractive index of the medium 1
n_2 = refractive index of the medium 2
θ_1 = angle of the incident ray from the normal (°)
θ_2 = angle of the refracted ray from the normal (°).

The same equation can be expressed in the terms of the refracted angle:

$$\theta_2 = \arcsin[(n_1/n_2) \sin \theta_1] \qquad (2\text{-}5)$$

4. The reflected flux for the randomly polarized incident wave can be calculated from the **Fresnel equation:**

$$\phi_r/\phi_i = \sin^2(\theta_1 - \theta_2)/2 \sin^2(\theta_1 + \theta_2) \\ + \tan^2(\theta_1 - \theta_2)/2 \tan^2(\theta_1 + \theta_2) \qquad (2\text{-}6)$$

In the case where the incident flux is perpendicular to the interface surface and the first medium is air, the equation reduces to

$$\phi_r/\phi_i = (n_2 - 1)/(n_2 + 1)^2 \qquad (2\text{-}7)$$

For a single surface or a single lens, as is most often used in optoelectronics application, the reflected flux is about 4% of the incident flux, a relatively small amount. In modern, more complex camera lenses with 10 or more elements, the reflection is a significant factor. Fortunately, reflection can be reduced to a tolerable level by applying to the lens surface a thin coating of material with a refractive indexes value between the index values of the two interfacing media.

EXAMPLE 2–2

A ray from the air is entering crown glass $n_2 = 1.52$ under an angle of $\theta_1 = 30°$. Find the ray angle in the glass and the ratio of the reflected and refracted flux to the incident flux.

Solution

Using designations in Figure 2–7 and Equation 2–5, we can write

$$\theta_2 = \arcsin[(n_1/n_2) \sin \theta_1] = \arcsin[(1/1.52) \sin 30°] = 19.47°$$

Using Equation 2–6 we can write

$$\phi_r/\phi_i = \sin^2(30 - 19.47)/2 \sin^2(30 + 19.47)$$
$$+ \tan^2(30 - 19.47)/2 \tan^2(30 + 19.47)$$
$$= 0.0334/1.155 + 0.0348/2.74 = 0.0415$$

So 4.15% of the incident flux is reflected; thus, 95.85% is refracted.

Going from less dense medium (with a smaller refractive index) to the higher density medium (with a greater refractive index), the ray always bends closer to the perpendicular. From the opposite direction, the ray bends away from the perpendicular, as seen in Figure 2–8. Snell's law, however, applies in either case.

As we can see from Figure 2–8, going from a denser medium to a less dense medium, as the angle of the incident ray increases, the exit ray bends

FIGURE 2–7   Reflection and refraction at $\Theta = 30°$ and $n_1 = 1.00$, $n_2 = 1.52$ interface.

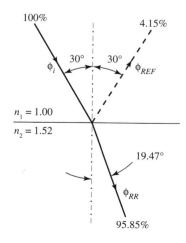

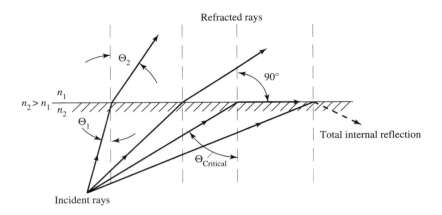

FIGURE 2–8   Internal reflection.

closer to the interface surface, and at some incident angle, called the **critical angle,** the exit ray is parallel to the interface surface ($\theta_2 = 90°$). Beyond that point the incident ray cannot pass through the interface surface and **complete internal reflection** takes place; that is, all incident flux is reflected back to the dense medium. The critical angle can be calculated from Snell's law, giving value to the exit angle $\theta_2 = 90°$:

$$\theta_c = \arcsin(n_1/n_2) \qquad (2\text{–}8)$$

where θ_c = critical angle (deg or rad).

Between crown glass ($n_2 = 1.52$) and air ($n_1 = 1$), the critical angle is 41.14°.

The phenomenon of total internal reflection has many uses in optical instrumentation, but its most significant application is in optoelectronics: the transmission of light in fiber optics is based on the practically lossless internal reflection in a transparent fiber. This application has opened an entirely new era of communication that we will discuss in greater detail in Chapter 8.

By far, the most important device based on the principle of refraction is the lens — a piece of circular transparent material with its faces polished into spherical surfaces. The lens is the most frequently used component in optical devices and has multiple optoelectronics applications.

Lenses can be separated into two general classes: **positive** or **converging** and **negative** or **diverging** lenses, as seen in Figure 2–9.

Positive lenses are the most useful for optoelectronics applications since they can concentrate the flux and form a real image of an object. Therefore emphasis in this text is on these lenses. (Negative lenses are used in complex lens systems, which are beyond the scope of this book.)

The properties of a given lens can be easily discovered by tracing rays through the lens, applying Snell's law at the entrance and exit surface of the lens. When the aim is to determine the **object** and **image relationship,** tracing only two of the three principal rays is necessary. In order to determine fine characteristics of a lens, the tracing of tens of thousands rays is required. This task falls to a lens designer, aided by a computer.

The **Newton Thin Lens Model,** which is an approximation, but very sufficient for our purpose, is used to delineate the object/image relationship. The basic terms and quantities are depicted in Figure 2–10. They are:

(a) The **optical axis,** which is an imaginary line through the center of the lens, perpendicular to the lens surface.

(b) The **nodal plane,** which is an imaginary plane at the center of the lens, perpendicular to the optical axis.

The bending of the rays takes place both at the front and rear surfaces of the lens — the ray is actually bent at two places. Newton's model makes a simplification and bends the ray only at one point, at the surface of the nodal plane. Distances to the focal points, object, and image are also measured from the nodal plane. More complex lenses with several elements, or thick lenses, have two nodal planes: the front and the rear nodal plane. Their locations are determined by the lens designer.

(c) The **focal point** *(F)* is a point on the optical axis where rays that are parallel to the optical axis converge, after passing through the lens. A lens has two focal points on either side of the lens.

(d) **Focal distance** or **focal length** *(f)* is the distance on the optical axis from the nodal plane to the focal point. The focal length is the most important characteristic of the lens since it describes lens "bending power."

FIGURE 2–9   Types of lenses.

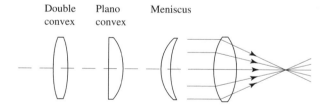

(a) Positive lenses

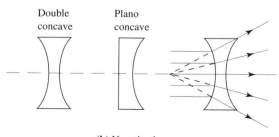

(b) Negative lenses

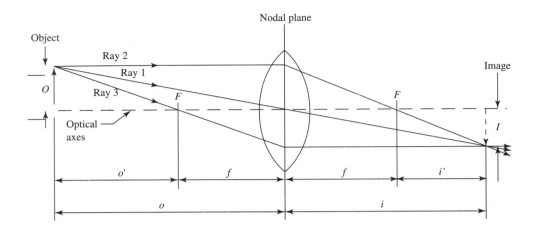

FIGURE 2–10   Ray tracing through a lens.

Short focal length lenses bend the rays more than long focal length lenses. When the medium on both sides of the lens is the same, the focal distances on both sides are the same also.

The **linear magnification factor** *(M)* is the ratio of image size to object size

$$M = I/O \qquad\qquad (2\text{–}9)$$

where M = magnification factor
  I = image size (m)
  O = object size (m).

When $M > 1$, the image is magnified; when $M < 1$, the image is reduced. In the latter case M is often called **reduction factor.**

2–3   DETERMINING OBJECT AND IMAGE LOCATION AND SIZE

When using lenses, the task of the designer is to control the flux from a source or to form an image of the source or object. Thus, it is necessary to determine object and image locations and their size relationship. Such determination can be accomplished in two ways: by graphical ray tracing or by using mathematical relationships.

The Graphic Ray Tracing Approach

Ray tracing uses simple rules for tracing three **principal rays** to determine object and image relationships. The principal rays and corresponding object and image distances are shown in Figure 2–10.

Principal ray 1 goes through the center of the lens. At that point, the lens surfaces are parallel and a ray going through a parallel plate does not bend but continues in the same direction with a slight parallel shift. Since the lens is considered thin, the shift is ignored. Principal ray 2 is parallel to the optical axis. After refracting through the lens, this ray passes through the focal point on the other side of the lens. Principal ray 3 passes through the focal point at the entrance side and emerges parallel to the axis after refracting through the lens.

At the intersection point of these three rays is the image of the object point where the rays emerged. Since the three rays all converge at the same image point, only two are needed to fix the point.

A graphic presentation, drawn in scale, is often quite satisfactory for solving lens problems in optoelectronics systems. When drawn to an enlarged scale, such a model may provide accuracy to within a few percent, sufficient in most cases. Also, it is the only method that reveals problems with obstruction in the ray path.

EXAMPLE 2–3

Find the image of a 4 mm diameter circle that is 15 mm away from a lens with focal length of 10 mm.

Solution
Figure 2–11 depicts the scale drawing of given conditions. Finding an image of the tip of the object circle radius allows us to construct the image circle, its size, and location.

The Mathematical Approach

The graphic method is the most satisfactory for solving lens problems in an existing system. For designing a new system, however, the mathematical approach is more efficient. In a new design the object, lens, and image positions

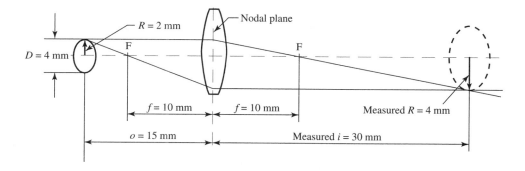

FIGURE 2–11   Graphic solution of Example 2–3.

and the lens focal length must be found from optical coupling specifications. They can be determined using the few simple relationships described below.

The focal length of a lens depends on the lens material (its refractive index) and the radii of the curvature of lens surfaces. It may be calculated using the **Lensmaker's equation:**

$$1/f = (n - 1)(1/R_1 + 1/R_2) \qquad (2\text{--}10)$$

where f = focal length (m)

n = refractive index of the lens material

R_1 = radius of the front face curvature (m)

R_2 = radius of the rear face curvature (m).

EXAMPLE 2–4

Find the focal length of a lens made from crown glass ($n = 1.52$) with a flat front face and a rear face radius (R_2) of 100 mm.

Solution
Using Equation 2–10 where $R_1 = \infty$, we have

$$1/f = (1.52 - 1)(1/\infty + 1/100) = 0.52 \times 0.01 = 0.0052 \quad \text{thus}$$

$$f = 1/0.0052 = 192.3 \text{ (mm)}$$

The prime lens equation can be determined from the geometry of similar triangles in Figure 2–10 with the following results:

$$1/f = 1/o + 1/i \qquad (2\text{--}11)$$

where f = focal length (m)

o = object distance from the nodal plane (m)

i = image distance from the focal plane (m).

Combining Equation 2–11 with the definition of the magnification factor, we can also use the form

$$M = I/O = i/o \qquad (2\text{--}12)$$

We can then develop a set of lens equations:

$$i = (of)/(o - f) \qquad (2\text{--}13)$$

$$i = f(1 + M) \qquad (2\text{--}14)$$

$$o = if/(i - f) \qquad (2\text{--}15)$$

$$o = f(1 + 1/M) \qquad (2\text{--}16)$$

$$o + i = f(2 + M + 1/M) \qquad (2\text{--}17)$$

$$f = (o + i)/(2 + M + 1/M) \qquad (2\text{--}18)$$

$$M = (o + i - 2f)/2f \pm \sqrt{(o + i - 2f)^2/4f^2 - 1} \qquad (2\text{--}19)$$

Equation 2–19 tells us that for a given object, image distance and focal length lens can have two positions for a sharp image: one that magnifies the image ($M > 1$) and another that reduces the image ($M < 1$). In this case, the magnification factor is the reciprocal of the reduction.

Besides the equations just presented, two more relationships may be of some value:

$$i' \times o' = f^2 \tag{2-20}$$

where i' = image distance from the focal point: $i' = i - f$ (m)
$\quad\quad o'$ = object distance from the focal point: $o' = o - f$ (m)

and

$$A_i/A_o = M^2 \tag{2-21}$$

where A_i = image area (m^2)
$\quad\quad A_o$ = object area (m^2).

Figure 2–10 shows that the image is a true enlarged or reduced replica of the object. However, it is upside down and perverted. In the case of photography, this state of affairs does not impose any problem since turning the negative right side up will correct the situation.

The object-image relationship can be described by three conditions, as seen in Figure 2–12:

(1) The object distance is more than two focal lengths ($\infty > o > 2f$). In this case the image distance is between one focal length and two focal lengths (an image never appears closer than one focal length) and the magnification factor M is less than one. This situation represents the case of a camera.

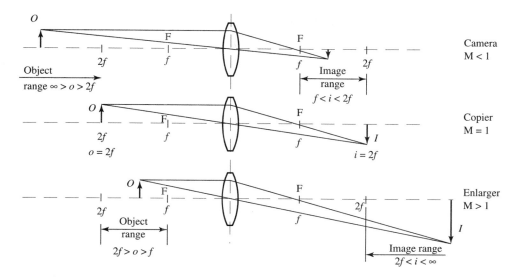

FIGURE 2-12   Three principal image-object conditions.

(2) The object distance is two focal lengths ($o = 2f$). In this case, the image distance is also two focal lengths and the image size equals the object size ($M = 1$). This situation represents the case of a same-size copier.

(3) The object distance is from two focal lengths to one focal length ($2f > o > f$). In this case, the image distance varies from two focal lengths to infinity and the magnification factor is always greater than one. This situation represents the case of an enlarger or projector.

When the object is closer than one focal length to the lens, no real image is formed.

Basically, the lens is used for two purposes: to form an image that can be read or detected by an electronic sensor (as in a video camera, for example), or to increase the optical transfer function. The latter is a simpler task and is one that an optoelectronics engineer encounters much more frequently.

The problem of very low optical coupling is well illustrated in Example 1–14 (see p. 37), where the OTF from a Lambertian source to a receiver 10 mm away was only 0.99%. The OTF can be improved by placing a lens between the source and the receiver, as shown in Example 2–5.

EXAMPLE 2–5

Improve OTF of the system in Example 1–14 by using a lens. Consider the source diameter to be 1 mm.

Solution

A lens is placed between the source and receiver in such a manner that source image fully covers receiver area. Figure 2–13 shows the geometry of the system.

The magnification in this case is

$$M = I/O = 2/1 = 2$$

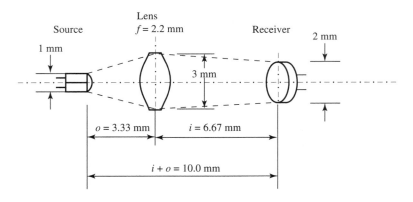

FIGURE 2–13   Example 2–5 coupling system.

Since the image-object distance is known, we can use Equation 2–19 to compute the lens focal length. Thus

$$f = (i + o)/(2 + m + 1/m) = 10/(2 + 2 + 0.5) = 2.22 \text{ mm}$$

and from Equations 2–14 and 2–15:

$$i = f(1 + M) = 2.22(1 + 2) = 6.67 \text{ mm} \quad \text{and}$$

$$o = f(1 + 1/M) = 2.22(1 + 1/2) = 3.33 \text{ mm}$$

Using a lens with diameter $D = 3.0$ mm placed in a calculated location, we can calculate the angle θ to the lens:

$$\theta = \arctan(D/2o) = \arctan 0.450 = 24.2° \quad \text{from here}$$

$$\text{NA} = \sin \theta = 0.411 \quad \text{and}$$

$$\text{OTF} = \text{NA}^2 = 0.169, \text{ or } 16.9\%$$

The addition of the lens provides an improvement of 17 times over direct coupling.

The use of a lens, however, brings another factor into the optical coupling that somewhat (but in most cases inconsequentially) decreases the flux transfer—some flux is absorbed when passing through the lens:

$$\phi_r = \phi_s \, \text{OTF} \, T \tag{2-22}$$

where T = transmission coefficient of the lens.

The **transmission coefficient** depends on the material and thickness of the lens. For most common optical glasses and lenses the coefficient is about 0.9 to 0.98 in the visible light region. For ultraviolet and infrared wavelengths, the coefficient may be considerably smaller, and special glasses or materials should be used. The transmission coefficient of a common lens material is shown in Figure 2–14.

Lens diameter has no effect on object-image position and size relationship. It does, however, affect image illumination. This property is described with the lens **F-number** *(F-NUM),* which is expressed as

$$F\text{-NUM} = f/D \tag{2-23}$$

where F-NUM = lens F-number
D = diameter of the lens opening (m).

The F-NUM is inversely proportional to the lens opening or aperture. Therefore a larger diameter lens that has greater illumination power has a smaller F-NUM.

Using F-NUM, the image illuminance and source luminance relationship is

$$E_i = \pi L_s T/4(F\text{-NUM})^2 \tag{2-24}$$

FIGURE 2–14   Transmission coefficient of optical glasses.

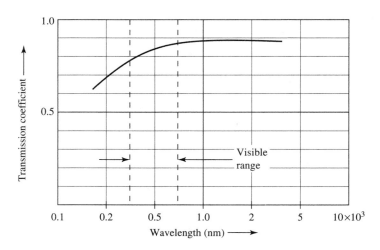

where E_i = image illuminance (W/m² or lm/m²)
L_s = source luminance (W/sr m² or cd/m²).

This equation applies when the image is close to the focal point, or for objects that are far away with $M \approx 0$. If that is not the case, a corrected equation has to be used:

$$E_i = \pi L_s T/4(F\text{-NUM})(1 + M)^2 \qquad \textbf{(2–25)}$$

where M = magnification factor.

SUMMARY

Chapter 2 introduces the two most common optical devices used in optoelectronics designs: the mirror and the lens. For mirrors, the basic rules of reflection, reflection coefficient, and image perversion are explained. Lenses are used to control and shape flux cones and to image objects. The text introduces elementary lens theory and develops a set of equations for determining object and image sizes and locations. Most optoelectronics problems involving a lens can be solved with these equations.

RECOMMENDED READING

Sears, F.W. and Zemansky, M.W. 1960. *College Physics.* Reading, MA: Addison-Wesley Publishing Co.
Smith, W.J. 1990. *Modern Optical Engineering.* New York: McGraw-Hill, Inc.
Meyer-Atent, J.R. 1984. *Introduction to Classical and Modern Optics.* Englewood Cliffs, NJ: Prentice-Hall, Inc.

PROBLEMS

1. An optical system operating in visible range uses two aluminum mirrors in the ray path. What percentage of flux is lost in the mirrors?
2. A human head (12 in. tall) is photographed with a camera with $f = 100$ mm lens from 6 ft. distance. Find image distance and size.
3. Source-to-receiver distance is 50 mm. The system requires that the source image be magnified 2 times at the receiver. Find lens focal length and location.
4. Find focal length of a lens made of crown glass $n = 1.50$ with front and rear curvature radii 15 cm and 25 cm.
5. A 3.00 mm diameter LED with $I_\theta = I_0 \cos^4\theta$ and $I_0 = 3.00$ mcd is coupled to a photodiode. The photodiode is 70 mm away and has a diameter 3.00 mm. What is the OTF and flux coupled into the photodiode when
   a. No lens is used between the source and receiver?
   b. A 10 mm diameter lens with transmission coefficient of 0.85 is used between the lens and receiver? Use $M = 1.00$.
6. The luminance of a frosted 60 W, 60 mm diameter incandescent bulb is 7000 cd/m². What is the illuminance of the bulb image projected through an $f = 50$ mm, F-NUM $= 1.4$, $T = 0.90$ lens which is 1 m from the bulb? Compare the results from Equations 2–24 and 2–25 and check result using flux calculation.
7. A 1 mm diameter Lambertian profile source is located 50 mm from a 2 mm diameter receiver. Calculate OTF when
   a. A 10 mm diameter lens which projects the source image to the receiver is used between the source and receiver.
   b. No lens is used.
8. Design an interrupting counter to count objects on moving production line. Given are:
   Source: A CaAs infrared (940 nm) source; diameter 1.5 mm, intensity $I_0 = 2.5$ mW/sr, pattern $I_\theta = I_0 \cos^n\theta$, $\theta_{1/2} = 15^0$.
   Receiver: Infrared photodiode; diameter 1.0 mm radiant responsivity 0.58 A/W at 940 nm.
      Use a 20 mm diameter lens, $T = 0.90$ that is located 70 mm from the source and project a 0.5 mm diameter image at the receiver. Design the system and find photodiode current when the beam is not interrupted.
9. Design a paper edge detecting system that detects the edge to an accuracy of ±0.5 mm. Given are:
   Source: as in Problem 2.8.
   Receiver: diameter 1.5 mm; responsivity 0.58 A/W.
      Use two lenses $f = 10$ mm, diameter 10 mm. Design a confocal system when the first lens projects a 1.00 mm diameter image at the paper edge plane, that is projected to the receiver.
10. What is the F-NUM of a 25.0 mm opening diameter, $f = 75$ mm lens?

The answers are in Appendix C.

3
Radiation Sources

Radiation sources, from candlelight and the warm glow of a fireplace to sunlight, are so common in our lives that we quite often overlook their presence and importance. The truth is that our entire existence depends upon radiation.

During the past few centuries many scientists devoted their talent and effort to explain radiation, especially the properties of light. Their combined efforts led to the successful development of the wave theory of light, which very accurately (and mostly in great detail) established the laws touched upon in Chapters 1 and 2. Wave theory has applications that extend to many complex natural phenomena.

One area, however, that completely defied explanation by the wave theory was the absorption of light by photosensitive materials and the generation of radiation and light. It took the postulation of quanta by Max Planck at the beginning of this century to explain those phenomena and to build a foundation for understanding elementary particles and the dual nature of energy and matter. For practitioners of optoelectronics who are involved only with *using* optoelectronics devices, and not *designing* them, understanding quantum mechanics (modern atomic particle theory) is not essential. However, some knowledge of the quantum theory is very helpful for understanding the operation of those devices. Therefore, an elementary review of quantum mechanics follows.

With the development of modern technology, new and more sophisticated optoelectronic sources have emerged to allow entirely new applications in communications, medicine, manufacturing technology, and so forth. In order to understand (and apply) these sources, it is necessary to describe and define their characteristics. We do so in the latter part of this chapter.

The approach to classifying and defining source characteristics is undertaken from the users' point of view so that they may be aided in selecting the correct sources along with the most effective practical adaptations for their selections.

Radiation sources can be divided into two groups: sources over which designers can exercise full control by selecting operational parameters and those where they cannot. Examples of controllable sources are incandescent bulbs, LEDs, and fluorescent panels. Sources such as CRTs and electroluminescent displays are much more complex devices and usually are produced in ready-made packages for sophisticated applications. The designer has very little control over the performance of these devices and is concerned only with their proper application. This book puts more emphasis on the first group of sources and describes the second only in terms of general principles and performance characteristics.

The remainder of this chapter is devoted to individual sources, their properties, and proper application practices. Understandably, the recent phenomenal growth in optoelectronics technology and the development of so many new sources makes explaining all existing devices in some depth virtually impossible. Consequently, information about the most common and useful sources is supplied.

3–1   THE ORIGINS OF RADIATION

To find the birthplace of radiation we must look very deep into the structure of matter — into the atom. The secrets of the atom, which were uncovered only at the beginning of this century, commenced with the investigation of radiation phenomena and resulted in a surprising theory of the dual nature of matter and energy.

During the investigation period scientists often developed **models of the atom,** to describe the common characteristics of matter. A very early model proposed by Niels Bohr, a famous Danish nuclear physicist, explains radiation quite satisfactorily.

According to Bohr's model, the atom consists of a tightly bound **nucleus** of positively charged **protons** and chargeless **neutrons** that are surrounded by **electrons** twirling in circular orbits around the nucleus. The electrons have extremely small mass, only 1/1847 the mass of a proton, and they carry a negative charge. Electrons are the particles that create an electronic current when moving from one atom to another. Figure 3–1 shows an atom based on the Bohr model.

Bohr was working with the hydrogen atom, the simplest atom of all, having only one proton and one electron. He tried to explain the absorption and radiation pattern of this gas: when heated, the gas emitted radiation in a particular set of distinct wavelengths. When light was passed through the gas, the same wavelengths were absorbed. The pattern is shown in Figure 3–2.

FIGURE 3–1   Bohr's model
of the atom.

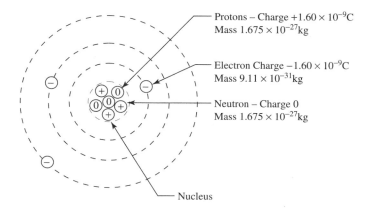

Protons – Charge $+1.60 \times 10^{-9}$C
Mass 1.675×10^{-27}kg

Electron Charge -1.60×10^{-9}C
Mass 9.11×10^{-31}kg

Neutron – Charge 0
Mass 1.675×10^{-27}kg

Nucleus

To explain this pattern, Bohr postulated the following conditions:

(a) The electron, instead of having only one orbit, may have several stable orbits. On the orbit closest to the nucleus, the electrons have the least energy; on the more distant orbits, the energy of the electron is greater. The radii of the orbits are characterized by the following relationship:

$$L = mvr_n = nh/2\pi \tag{3–1}$$

where L = the angular momentum of the particle (kgm^2/s)
    n = integer; $n = 1, 2, 3 \ldots$.
    h = Planck's constant 6.626×10^{-34} (Js)
    m = electron mass (kg)
    r = radius of the orbit (m).

The purpose of presenting this equation is not to teach how to calculate the fundamental parameters of an atom but to point out the significant role of Planck's constant in every atomic relationship.

Using the laws of mechanics and Coulomb's law of attraction of charges, the radii can be calculated. The closest radius, $r_1 = 0.529 \times 10^{-10}$ m, and the other radii are connected to r_1 with relationship $r_n = r_1 n^2$. Therefore, the set of possible orbits for a hydrogen atom electron is shown in Figure 3–3.

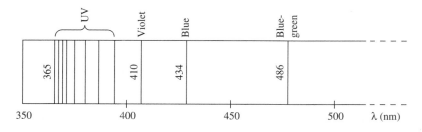

FIGURE 3–2   Radiation spectrum from hydrogen (Balmer series).

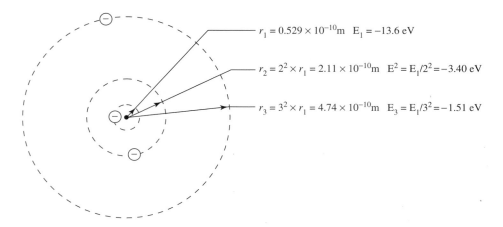

$r_1 = 0.529 \times 10^{-10}\text{m}\quad E_1 = -13.6\ \text{eV}$

$r_2 = 2^2 \times r_1 = 2.11 \times 10^{-10}\text{m}\quad E^2 = E_1/2^2 = -3.40\ \text{eV}$

$r_3 = 3^2 \times r_1 = 4.74 \times 10^{-10}\text{m}\quad E_3 = E_1/3^2 = -1.51\ \text{eV}$

FIGURE 3–3   Electron orbits and energy levels of hydrogen atom.

(b) To reach higher orbits, electrons require energy, a situation similar to an orbiting space shuttle that requires more fuel to reach a higher orbit. The energy level can be expressed in joules (J), but as is customary in atomic physics, it is expressed in eV, **electron volts,** which is the amount of energy required to transmit the charge of an electron across the potential difference of 1 V. Its relationship to a joule is

$$1\ \text{eV} = 1.602 \times 10^{-19}\ \text{J} \qquad \textbf{(3–2)}$$

The energy characteristics of an atom are usually presented in the form of an **energy-level diagram,** as shown in Figure 3–4.

The diagram shows that the lowest energy level is called the **ground state** ($n = 1$, the lowest orbit). The higher orbits have higher energy levels. However, there is a limit of energy an electron can achieve before the electron escapes from the atom. This limit is called the **maximum energy level** ($E_{\max}$). It is customary to assign the value zero to $E_{\max}$. Consequently, since the other orbits all have a lower energy level, they assume negative values.

The ground state ($n = 1$) is the most natural state of the electron. However, by exciting the atom with heat, radiation, or by many other means, the electrons can absorb additional energy and thus move to a higher orbit. In this case, the atom is in an excited state. When the energy added to the electron exceeds the maximum energy level, or binding energy, the electron breaks the atomic bonds and escapes, leaving the atom one electron short. The positive charge of the proton is now no longer balanced by the negative charge of the electron and a positive ion is formed.

(c) The space shuttle that can be boosted to a higher orbit by adding the energy through burning more fuel can also be brought to a lower orbit. In this case, the energy of the shuttle must be dissipated. (This, incidentally, was one of the most difficult design problems of the shuttle — how to dissipate at re-entry the huge amount of energy stored in the shuttle.)

The same condition exists with orbiting electrons when their energy must be released. In the case of an electron, energy is dissipated through a form of radiation: an electron descending from a higher energy level to a lower one emits radiation.

There is, however, a significant difference between a space shuttle and an electron — the shuttle can move to any arbitrary energy level or orbit. Not so with electrons. They can exist only on predetermined orbits and energy levels. Therefore, the release of radiation from atoms is also in predetermined quanta, called **photons**. According to Einstein's equation, the energy of the photon is

$$E = hf \tag{3-3}$$

where E = photon's energy (J)

f = frequency of radiation (Hz)

h = Planck's constant; h = 6.626×10^{-34} (Js)

The notion that the energy of radiation is generated in definite packets provides the basis for the theory that light and radiation have a "granular"

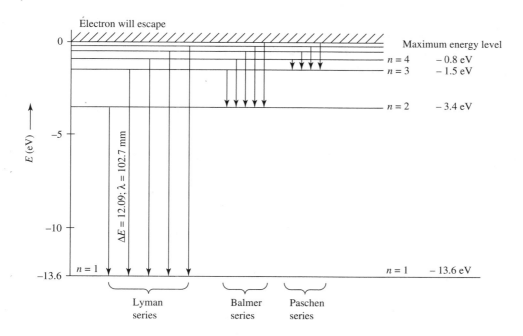

FIGURE 3–4   Energy levels and photon wavelength in a hydrogen atom.

nature and react in certain conditions as particles and in other conditions as waves. This concept has found further proof in later theories and experiments. This seemingly dual character of light is a most interesting and perplexing manifestation of nature and forms part of the foundation of modern atomic physics.

The photon — defined as a massless particle traveling at the speed of light — has an energy level determined by Einstein's equation, and can react with electrons within the atom's structure. When a photon is absorbed by an atom, an electron in the atom is elevated to a higher orbit or energy level; when a photon is emitted, an electron is dropped to a lower energy level.

The best proof of this postulation is found in Bohr's model of the hydrogen atom. The differences in the energy levels in Bohr's atom can be converted to corresponding radiation quantas or photons, using his theory.

The energy level of any orbit can be calculated from the ground state level:

$$E_n = E_0/n^2 \quad \text{or} \quad E_n = -13.6/n^2 \tag{3-4}$$

where E_n = energy level at the n-th orbit (eV)
E_0 = ground energy level $E_0 = -13.6$ (eV)
n = orbit number.

EXAMPLE 3–1

Find the energy level of the third orbit and the energy loss when an electron drops from the third orbit to the ground state.

Solution
The energy level of the third orbit is

$$E_3 = -13.6/3^2 = -1.51 \text{ eV}$$

Energy released when an electron falls from third orbit to the ground state is

$$\Delta E = E_3 - E_0 = -1.51 - (-13.6) = 12.09 \text{ eV}$$

The released energy is transferred into a photon, the frequency of which can be calculated according to Equation 3–3 (Einstein's equation) as

$$\Delta E = hf$$

When the frequency is converted into wavelength and electron-volts are converted into joules, the expression for photon wavelength can be developed:

$$\lambda = hc/(E_n - E_{n'}) \tag{3-5}$$

or, when converted to electron volts

$$\lambda = (6.63 \times 10^{-34})(3.00 \times 10^8)/(E_n - E_{n'}) \times 1.602 \times 10^{-19}$$

$$= 12.42 \times 10^{-7}/(E_n - E_{n'}) \tag{3–6}$$

where λ = photon's wavelength (m)

$(E_n - E_{n'})$ = energy difference between levels of n and n' (eV).

Carrying this calculation out for third and ground levels, where the energy difference was 12.09 eV, we obtain the wavelength

$$\lambda_{3-0} = 12.42 \times 10^{-7}/12.09 = 1.027 \times 10^{-7} \text{ m} = 102.7 \text{ nm}$$

This wavelength corresponds to a line in the **hydrogen radiation spectra** (second line in the Lyman series). Similar calculations are carried out for several other energy shifts and corresponding radiation wavelengths. The results are shown in Figure 3–4. Comparing the results to Figure 3–2 (hydrogen radiation spectra) gives a very good correlation of measured and calculated data. This is convincing proof of the correctness of photon and radiation theory.

Bohr's atom model is one of the earliest explanations of atomic structure and radiation. In more complex atoms with many electrons, the theory is quite complex and beyond our present scope. The basic principle that photon emission or absorption is connected to a change in an electron's energy level always applies. Therefore, it can be deduced that in atoms with many electrons and possible energy levels an almost continuous spectra of radiation can be observed which, in fact, is the case of radiation from heated metals.

Another radiation theory, called **indirect transition,** illustrated in Figure 3–5, has important applications for optoelectronic devices.

As we discussed, a hydrogen atom can absorb a photon of 12.1 eV energy, or 102.7 nm wavelength (in ultraviolet region), which moves the electron to the third energy level. There is a possibility that when the electron falls back to the ground state, it may happen in two steps: first falling back to the second level and from there to the ground state. In this case, two photons are

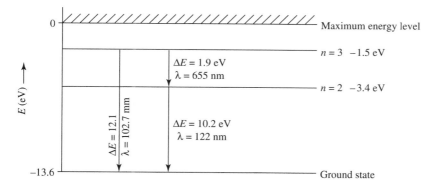

FIGURE 3–5   Indirect transition of an electron.

emitted, the first one with an energy level of 1.9 eV, which corresponds to the wavelength 655 nm (dark red light), and the other with an energy level of 10.2 eV, or 122 nm (ultraviolet). The total energy of emitted photons is equal to the energy of the absorbed photons, but a very interesting phenomenon is taking place: by radiating an atom with photons at a higher energy level, a lower level photon or longer wavelength radiation is emitted.

This effect, called **fluorescence,** has significant applications in the opto-electronics field. The best known is the fluorescent light bulb; ultraviolet radiation emitted by the electrically excited gas inside the bulb is transformed into visible light through the fluorescence of the phosphorous coating on the inside surface of the bulb.

One important application of excited radiation is the radiation from semiconductor materials (LEDs). A short overview of this phenomenon is given next.

Semiconductor Junction

The semiconductors, silicon (Si), germanium (Ge), gallium (Ga), etc., have much more complex atoms than the hydrogen just described. Silicon, for example, has 28 protons and electrons as compared to only one of each in hydrogen. In semiconductor atoms, the electrons are organized into **shells** designated by letters K, L, M, and N. Each shell has a fixed number of electrons: the K-shell has 2, the L-shell has 8, the M-shell has 18, and the N-shell has 32. All shells must be filled, with the exception of the outermost shell, which takes the electrons left over from other shells.

For example, germanium has 28 electrons that fill shells K, L, and M with 4 electrons left over in the N-shell, as seen in Figure 3–6.

FIGURE 3–6   Shells in a germanium atom.

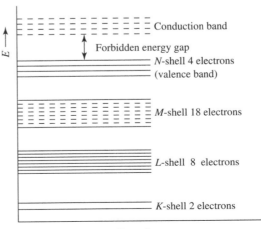

Ground state

The last shell is called the **valence band,** and the electrons there, the **valence electrons.** Those electrons engage in chemical reactions with other elements and determine the electrical characteristics of the element. By applying additional energy, the electrons can be lifted to a higher energy level by crossing the **forbidden energy gap** into the **conduction band.** In this band, electrons are free to follow electrical forces and travel from one atom to the other. They become the carriers of electricity.

The electrical properties of a material depend upon the availability of carriers and the number of electrons in the conduction band. The conductors may have as many as 10^{23} carriers per cm^3. A good insulator may have as few as 10 carriers per cm^3. Semiconductors fall between those two, with 10^8 to 10^{14} carriers per cm^3. Also, there is a difference in the forbidden energy gap. In conductors, the gap is practically nonexistent, but in insulators the gap is large. Again, the semiconductors are in between.

The conductivity of a semiconductor can be increased in two ways. The first is by applying thermal energy, or heating the semiconductor, lifting more electrons into the conduction band. This method is effective with all semiconductors, but in general it is not desired because it makes conductivity dependent on environmental conditions. The second method for increasing conductivity is to introduce free electrons by doping the semiconductor, that is, by adding materials having excess carriers, atoms that introduce either negative ions (with excess electrons) or positive ions (with shortage of an electron). This method gives to the semiconductor designer control of semiconductor conductivity.

As seen from Figure 3–7(a), all valence electrons are engaged with neighboring atoms, leaving no free electrons for the conduction band. Therefore, the conductivity for this type of a material is low.

Figure 3–7(b) shows a semiconductor doped with an atom (arsenic) having 5 protons and 5 electrons. The fifth electron will not fit into a normal 4-electron valence band and is therefore free to move from one atom to another, and to carry electricity. Since the carrier is an electron, a negatively charged particle, this type of semiconductor is called an **n-type semiconductor.**

Figure 3–7(c) shows a semiconductor doped with an atom (gallium) having 3 protons and 3 electrons. In this case, one space between neighboring atoms is left empty — there is a hole. This atom is now a positive ion, which likes to attract an electron from the next atom. By doing this, the hole is moved to the next atom; there is a movement of holes and electrons, in opposite directions, but an electric charge is still carried and the conductivity of the semiconductor is increased. Since the charge movement is caused by **holes,** or positively charged atoms, this type of semiconductor is called **p-type.**

The energy levels of n- and p-type semiconductors are illustrated in Figure 3–8.

Figure 3–8(a) shows energy levels in an n-type semiconductor. Their high number of electrons is close to the conduction band. The electrons have a higher energy level and thus the average energy level of n-type semiconductors is higher than in pure semiconductors.

FIGURE 3–7   *n*- and *p*-type semiconductors.

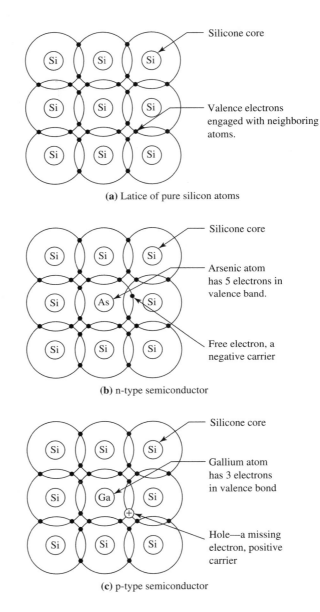

(**a**) Latice of pure silicon atoms

(**b**) n-type semiconductor

(**c**) p-type semiconductor

In *p*-type semiconductors fewer electrons are close to the conduction band and thus the average energy level is lower than the energy level of a pure semiconductor, as shown in Figure 3–8(b).

The *p*-type material has a large number of holes that are free to move and carry current. These holes are called the **majority carriers.** In *n*-type materials the electrons are the majority carriers and the holes are the

minority carriers. They both carry current and move in opposite directions, but the majority carriers are the predominant current carriers.

Radiation from a Semiconductor Junction

From the standpoint of optoelectronics, the most important radiation source is the semiconductor junction, commonly called the **p-n junction.** In a *p-n* junction, a *p*-type semiconductor and an *n*-type semiconductor are fused together. This junction has many interesting properties, among them the ability to emit radiation.

At *p-n* junctions the free carriers—the electrons and holes—diffuse toward one another and combine, leaving on both sides of the junction a **depletion region** that is nonconducting since there are no longer any free carriers. On the *n*-type side of the depletion region, where the electrons diffused to *p* side, is a collection of positively charged ions that forms a potential barrier to prevent further diffusion of electrons. On the *p* side, the barrier is negative since the holes in the atoms were filled. Thus, a **barrier voltage** exists across the junction.

When a reverse bias is applied to the junction, it adds to the barrier voltage. As a result, the depletion region is increased and no current will pass through the junction. When a forward bias is applied, the bias counteracts the barrier voltage, and the depletion region is decreased. When the bias exceeds the barrier voltage, the electrons and holes combine, and current will flow. In this process, since the electrons in the conduction band are at a higher energy level than the holes in the valence band [as seen in Figure 3–9(a)], some energy in the form of radiation is released. The phenomenon is called **electroluminescence** when the wavelength emitted is in the visible range.

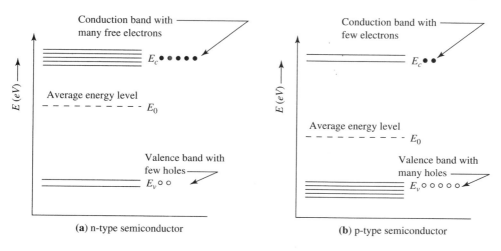

(a) n-type semiconductor   (b) p-type semiconductor

FIGURE 3–8   Energy levels in *p-n* junction.

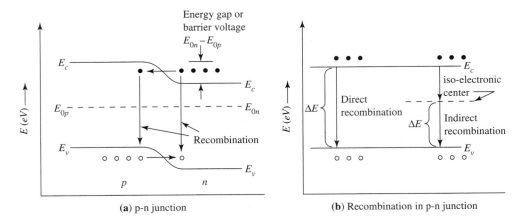

(a) p-n junction   (b) Recombination in p-n junction

FIGURE 3–9   Recombination in a *p-n* junction.

The wavelength of emitted radiation depends upon several factors. The first, the energy gap or barrier voltage between the *p* and *n* energy levels, is determined by the semiconductor material. The second is the mode of recombination. It can be direct [Figure 3–9(b)], when the electron drops directly from the conduction band to the valence band, or indirect, when the electron is first trapped in a neutral **iso-electronic** center. The mode of recombination depends upon the semiconductor material and its impurities.

In either mode of recombination the wavelength of radiation depends upon the energy gap or barrier voltage and can be calculated from the following equation:

$$\lambda = 1240/\Delta E \qquad \qquad \textbf{(3–7)}$$

where λ = wavelength of radiation (nm)
   ΔE = energy gap (eV).

EXAMPLE 3–2

GaAs semiconductors have an energy gap of 1.43 volts. Find the radiation wavelength.

Solution
Using Equation 3–7 we have

Δ = 1240/1.43 = 867 nm, which is in the infrared region.

The most common semiconductor materials, silicon and germanium, have relatively low energy gaps (1.09 and 0.66 eV, respectively) and therefore emit radiation at the long wavelengths of 1140 and 1880 nm, which are in the infrared region. Semiconductors with higher energy gaps are gallium (Ga), arsenic (As), phosphorus (P), indium (In), and antimony (Sb). They radiate at shorter wavelengths. Figure 3–10 shows the wavelength of emissions of those semiconductors. Also in the same figure, the corresponding energy gap, the range of human vision, and the sensitivity range of silicon PIN diode are shown.

3–2   CLASSIFICATION OF RADIATION SOURCES

The following section classifies optoelectronic sources by their primary characteristics. The classification should aid in the selection of sources for specific applications.

Classification by the Flux Output

The differences in flux distribution from the most common types of sources are shown in Figure 3–11.

A **point source** [Figure 3–11(a)], typically an LED or a small filament clear bulb, has a small emission area and is the most common radiation source. The incidence and the illuminance are inversely proportional to the distance squared from the source. The relationships defined in Chapter 1, pp. 32–35, apply.

An **area source** [Figure 3–11(b)], typically an electroluminescence panel or frosted light bulb, has an emission area that is large when compared to its distance from the observer. Relationships defined in Chapter 1, pp. 42–45, apply to this condition.

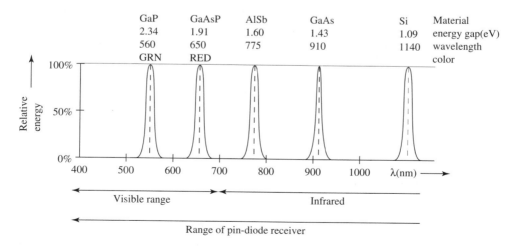

FIGURE 3–10   Semiconductor's radiation sources.

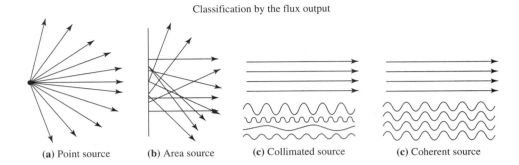

Classification by the flux output

(a) Point source   (b) Area source   (c) Collimated source   (c) Coherent source

FIGURE 3–11   Flux lines from different sources.

A **collimated source** [Figure 3–11(c)], such as a searchlight, has flux lines that are parallel, and therefore, the square law relationship with distance and illumination does not apply.

A **coherent source** [Figure 3–11(d)], such as a laser, is either a point source or a collimated source with one important difference: the waves in a coherent source are all in phase.

Classification by Wavelength and Color

The wavelength, or color when the radiation is in the visible range, is certainly one of the most important characteristics of a source. Three characteristics define the color:

(1) **Hue** is what is commonly meant by the term "color." Blue, red, magenta, etc., designate the hues. The wavelength of radiation determines the hue, but as explained later, wavelength is not the only factor.

(2) **Saturation,** or purity as it is also called, is another color characteristic. Colors in nature are not pure, single wavelengths, but contain varied amounts of white light, decreasing the saturation of the color. Pink, for example, is a mixture of red and white. Thus, it is red with less saturation. Pure white has zero saturation and therefore "dilutes" or decreases saturation when added to any color.

(3) **Intensity** describes the flux density of a radiating source or luminance of a reflecting surface.

Very often the term **chroma** is used in connection with color phenomena. The chroma is the combination of hue and saturation. Therefore, a color in technical literature is described by two parameters: chroma and intensity.

Since colors are perceived through the human eye and the brain, their characteristics must be considered when dealing with color and color perception. We perceive the different wavelengths between 400 and 700 nm as different colors. The wavelength of 490 nm, for example, is perceived as green.

However, it is peculiar to our vision that a mixture of blue (460 nm) and yellow (580 nm) gives us the same perception, and we cannot distinguish which green is a "pure" green or which is a mixture. In this respect, the visual response is different from hearing. When we hear a chord, the ear can distinguish the component tones in the chord. The eye does not have this capability.

This property of the eye allows us to create new colors by mixing two or more colors. In fact, we need only three properly selected colors to generate all possible hues and their saturations. Those colors are called **primary colors.** C.I.E. has selected three monochromatic colors as primary colors: blue (440 nm), green (546 nm), and red (700 nm).

Any color could be a mixture of three primary components or **tristimulus values,** as they are called. Any chroma can be specified by three quantitative values: $C = B + G + R$ where B, G, and R are the primary flux components in the flux of color C. This relationship could also be written in terms of the relative proportions of x, y, and z in a color mix, where $x = B/(B + C + R)$, $y = G/(B + G + R)$, and $z = R/(B + G + R)$. These three values could also be used to specify a color. Because $x + y + z = 1$, there is a redundancy in using all three parameters and thus only two, x and y, are needed to specify a color or chroma. The letter Y is used to specify the intensity or luminance of the source.

Using these parameters, or coordinates as named by C.I.E., a standard C.I.E. chromaticity diagram may be constructed (see Figure 3–12).

In this figure, the horseshoe-like curve from points R to B is called the **spectrum locus.** All saturated hues that exist in the rainbow can be represented as monochromatic hues that lie on this curve and can be identified by a corresponding wavelength marked on the curve. The straight line from R to B represents the line of purples. The purple hues that cannot be created by a single wavelength, but only by mixture of red and blue, do not appear in the rainbow colors. Point A in the center of the curve, with coordinates $x = 0.33$ and $y = 0.33$, represents the **achromatic point,** a pure white without any hue, or a point with zero saturation. A line drawn from this point to a given hue on the spectrum locus represents all possible saturations of this hue. A line from a point on the spectrum locus through the achromatic point A to the other side of the spectrum locus identifies the **complementary color** of the first point.

Using the C.I.E. chromatic diagram and an instrument called a colorimeter, any color can be measured and specified. This ability to quantify is quite important when specifying the color of optoelectronic sources. Colors of most common sources are located in Figure 3–12.

The C.I.E. chromatic diagram is also helpful in understanding the term **color temperature,** which describes the hue of very low saturation sources or, in other words, the hue of almost white light.

Color temperature is mostly used in connection with sources that produce a continuous spectrum of light, such as an incandescent bulb. This type of spectrum is generated by lifting the electrons to a higher energy level by

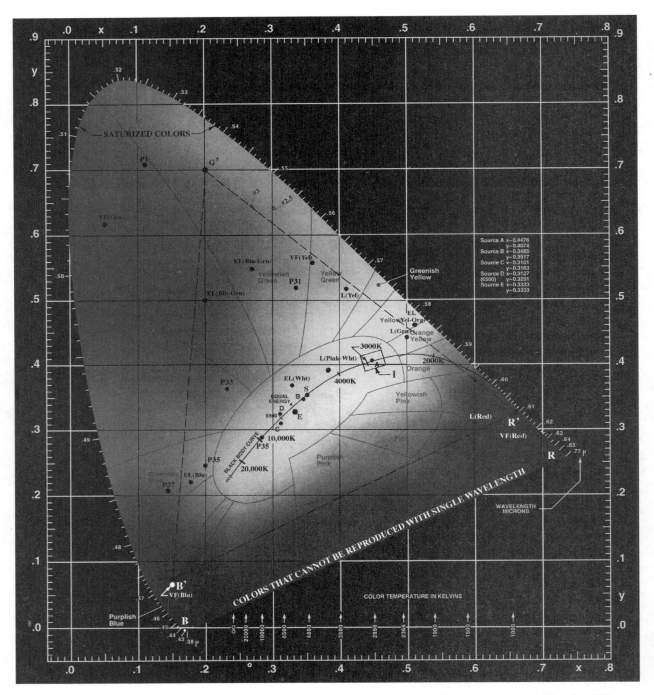

The markings on the diagram are: E—equal energy point (pure white); S—sunlight; R', G', B'—color TV primary colors; P—CRT phosphors; L—LED sources; VF—vacuum fluorescent sources; EL—electro-luminescent panels.

FIGURE 3–12 C.I.E. chromatic diagram with loci of common source colors. A color version of this diagram is on the back cover. (Courtesy Photo Research Corp.)

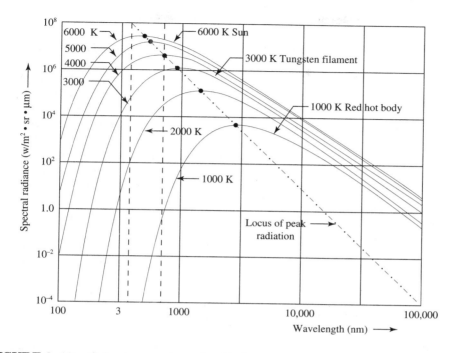

FIGURE 3–13   Color temperate and radiation spectrum of blackbody radiator.

thermal energy; that is, by heating a material, usually metal. By applying more energy or heating the object to a higher temperature, the radiation pattern from the object changes. A typical pattern is shown in the well-known black-body radiation curves in Figure 3–13.

As seen from the curves, when the radiator temperature is increased, two things happen. First, the intensity of the radiation increases and the spectrum shifts toward the lower wavelengths. When we look at the spectrum distribution in the visible range, we can see that a source with a temperature of 1800 K, such as candlelight, has a radiation distribution only in reds and yellows. Therefore, the light from a candle looks reddish. A 500 W incandescent lamp radiates over the entire visual spectrum with more energy in the red and yellow wavelengths and therefore this light is relatively yellow. Sunlight has more energy in blues and greens and produces a bluish light. The term color temperature is used to identify such distribution of the spectrum. The same thing can also be shown on the C.I.E. chromaticity diagram[1] where a line starting from a spectral red at 610 nm indicates the locus of all chroma corresponding to a color temperature.

---

[1]It should be noted that the chromatic diagram described here is the 1930 version of the C.I.E. diagram. In 1970, C.I.E. developed a new version of the diagram with a different coordinate system. The 1930 version is employed in this text because it is still widely used and is somewhat easier to explain. The conversion of chroma coordinates from one version to the other can easily be done using equations given with the 1970 diagram chart.

Now, with some understanding of color at hand, we can classify the sources by radiation spectrum and color, as shown in Figure 3–14.

A **continuous spectrum** source is usually a thermally excited source and has a wavelength of emission that ranges from ultraviolet to infrared. The spectral distribution in the visible range may be described by chroma coordinates or color temperature, the latter being the more popular method. Incandescent bulbs are typical representatives of this kind of source.

A **line spectrum** source has distinct narrow bands of radiation throughout the ultraviolet to infrared range. Gas discharge and phosphorescent sources have this type of radiation pattern. A typical example is a fluorescent lamp where argon-neon mixture, with traces of mercury, is excited by an applied current to produce the radiation. The line spectrum emitted from this source is mostly in the ultraviolet range. This flux excites the phosphor coating on the bulb surface, converting the ultraviolet energy to a line spectrum in the visible range.

The sources are specified by giving either the wavelength spectrum or the combined color in the visible range, using the color temperature or chroma coordinates.

A **single wavelength** source radiates only in a narrow band of wavelengths. Electroluminescent sources, LEDs, and electroluminescent panels

FIGURE 3–14   Radiation spectras of sources.

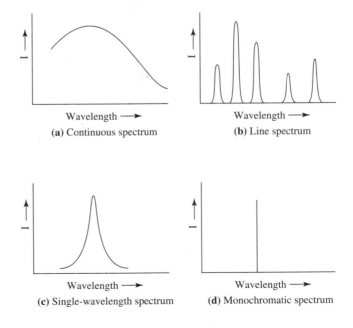

(a) Continuous spectrum

(b) Line spectrum

(c) Single-wavelength spectrum

(d) Monochromatic spectrum

are typical examples of this type of source. The source is specified by radiation frequency, color, or chroma coordinates.

A **monochromatic source** radiates at a single wavelength or in a very narrow band of wavelengths. The laser is a typical example of such a source. It is usually specified by the wavelength.

Classification by the Source Excitation

As we have discussed, to emit radiation electrons must be elevated to a higher energy stage. Therefore, the excitation of a radiating medium is necessary to produce radiation. The method of excitation and resulting radiation characteristic depend on the media (material) of the source. The method of excitation, however, is the prime factor in determining radiator characteristics. Following are the most common, but by far, not all modes of excitations:

Thermal excitation, heating the material, is the most common mode of excitation. It produces a continuous spectrum of radiation when used with solids and a line spectrum with gases. Incandescent filament is a typical representative of this method.

Electroluminescence is the mode of excitation in semiconductors (see p. 95). It produces a narrow band, single-wavelength spectrum. It is the mode used in LEDs and related devices.

Vacuum fluorescence uses a high energy electron beam in a vacuum. When the beam strikes the medium, or target, the beam energy is transferred to the electrons of the medium, and upon their decay, radiation occurs. Depending on the variety of electron-beam energy and target material, the radiation is single wavelength with a frequency from visible to X-rays. Cathode ray tubes, TV tubes, and X-ray machines are typical representatives of this excitation mode.

Chato-luminescence takes place around the cathode electrode of a low pressure gas-filled tube. When current is passing through the tube, a high voltage gradient around the cathode elevates gas electrons to an excited state and introduces glow in this region. This mode is the operating principle of plasma and gas discharge devices.

Lasing action can take place in many optical mediums with proper systems that include optical feedback. Numerous exciting methods are used. Most common are optical, gas discharge, electroluminescence, and chemical.

More detailed discussion of exciting modes is included in Chapter 4, pp. 141–151. The exciting mode provides a convenient and logical way to group and classify most of the sources. (See Table 3–1.)

The tables bring a general overview and guideline of most common optoelectronic sources. They will help to match the source characteristics to possible applications.

TABLE 3–1
Listing of Sources by Source Excitation Mode

Excitation	Thermal	Electric Current in Semiconductors			Electron Beam in Vacuum		Electric Current in Gas		Optical Gas Discharge Chemical
Luminescent Phenomenon	Incandescence	Electroluminescence			Vacuum Fluorescence		Chato-Luminescence		Laser
Types of Devices	Light Bulbs 7-Segment Displays	LED	7-Segment Dot Matrix Displays	Luminescent Panels	Vacuum Fluorescent Displays	CRT	Gas Discharge Lamps	Plasma Displays	Communication, Industrial, Military, Medical, etc.
Working Voltage	M,H	L,M	L	H AC	M	H	H	H	H
Flux Output	Point Source Area Source	Point Source	—	Area Source	Area Source	Point Source	Point Source	Area Source	Point-Source Coherent
Intensity	M,H	L	L	M	M	M	M	M	M,H
Spectrum	Continuous	Single-Frequency Wide Band			Single-Frequency Wide Band		Single-Frequency Wide Band		Monochromatic
Cost	L,M,H	L	M	M	M	H	L	H	H
Life	M,H	H	H	M,H	H	H	H	H	H

TABLE 3-2
List of Arbitrarily Chosen Ranges of Characteristics

Characteristic	Low Range (L)	Medium Range (M)	High Range (H)	Units
Working voltage	<1	1–30	>30	V
Flux output	<1	1–100	>100	W, lm
Intensity	<1	1–100	>100	cd
Life	<100	100–3000	>3000	h
Cost	<0.3	0.3–30	>30	$

3-3   RADIATION SOURCES

The following section describes the most common optoelectronics sources, their main characteristics, and design and application rules.

Incandescent Source

An incandescent source, the light bulb, is by far the most common optoelectronics source. In the last few decades its use in electronic instrumentation and application has greatly diminished, mostly because of its high power consumption as compared with semiconductor circuits. Nevertheless, light bulbs still are used for pilot lights, annunciators, panel illuminators, daylight readable 7-segment displays, etc.

The mechanical construction of a light bulb varies quite a lot, as can be seen in any bulb catalog. The electrical characteristics of most light bulbs, however, are similar. They present a compromise between efficacy, light or flux output, color temperature, and life. The bulb designer selects the compromise, but very often it is advantageous to modify those characteristics to meet special use requirements. Increasing the pilot light life in situations where accessibility to the bulb is difficult, is a typical example.

The relationships among the bulb life, light intensity, efficacy, and current as a function of applied voltage are given in following equations.

The intensity:

$$I_1 = I_0 \times (V_1/V_0)^{3.5} \qquad \textbf{(3-8)}$$

where V_0 = nominal bulb voltage (V)
  V_1 = applied bulb voltage (V)
  I_0 = nominal intensity (cd)
  I_1 = intensity at voltage V_1 (cd).

Since the radiation pattern does not change with changing voltage, the same equation applies to flux also.

Bulb life:

$$L_1 = L_0 \times (V_0/V_1)^{12} \qquad \textbf{(3-9)}$$

where L_0 = nominal bulb life (h)
L_1 = life at applied voltage V_1 (h).

Bulb current:

$$i_1 = i_0 \times (V_1/V_0)^{0.55} \qquad\qquad \textbf{(3–10)}$$

where i_0 = nominal bulb current (A)
i_1 = current at voltage V_1 (A).

Efficacy:

$$K_1 = K_0 \times (V_1/V_0)^{1.9} \qquad\qquad \textbf{(3–11)}$$

where K_0 = nominal bulb efficacy (lm/W)
K_1 = efficacy at voltage V_1 (lm/W).

These relationships in graph form are shown in Figure 3–15.

Another factor to consider when using an incandescent bulb is the **inrush current.** The bulb is a nonlinear resistor. Because of the temperature

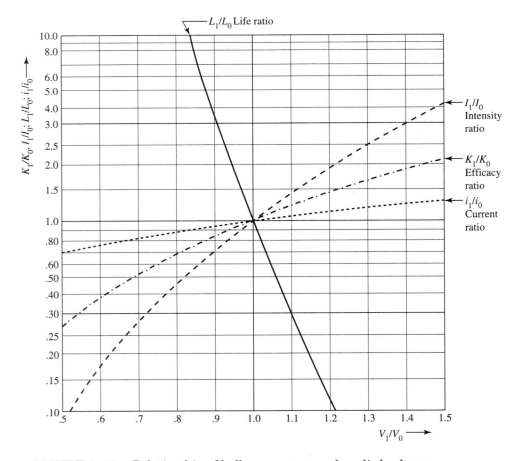

FIGURE 3–15   Relationship of bulb parameters and applied voltage.

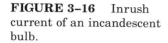

FIGURE 3–16 Inrush current of an incandescent bulb.

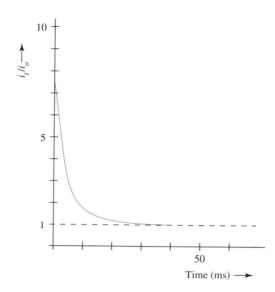

coefficient of the tungsten filament, the resistance of the hot filament is about 8 to 10 times higher than when it is cold. Therefore, the bulb, for a very short period when it is first turned on, draws a higher current. Thus the term "inrush current." A typical inrush current graph is shown in Figure 3–16.

This condition has considerable significance for high incandescent loads — special switches and contactors must be used to prevent contact damage. For small pilot lights and similar applications the amount of inrush current has no significance, except when a transistor or semiconductor device is used to drive the bulb since its rating has to withstand the inrush current.

Lastly, a note about color temperature of the light from an incandescent source: it is determined by the filament temperature and is a function of applied current. A typical relationship between color temperature and bulb efficiency is shown in Figure 3–17. The **efficiency** is defined as the ratio of bulb flux (lm) to total input power (W).

With the advance of semiconductor circuitry, the use of incandescent bulbs has diminished, mainly because of their relatively high power consumption and large size. However, their high flux output and white light, which with the use of different color lenses can be converted to any color, make them popular as panel indicators. They also work well in 7-segment displays where daylight readability is required. A few typical examples of such devices are shown in Figure 3–18. They can be driven with AC, be multiplexed, and are TTL compatible.

EXAMPLE 3–3

A Radion type ML/7216 lamp with specifications $V_0 = 5.0$ V, $i_0 = 0.125$ A, $I_0 = 0.22$ cd, and $L_0 = 5000$ hours is intended for use as a pilot light in

FIGURE 3–17   Relationship between bulb efficiency and color temperature.

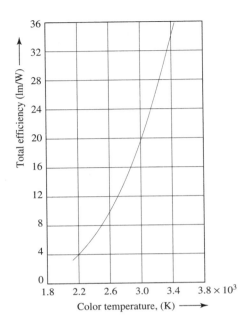

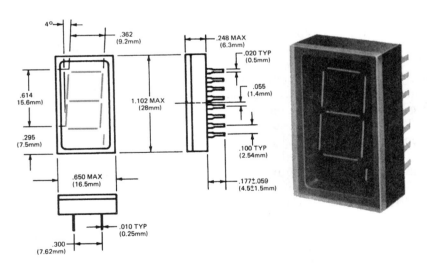

FIGURE 3–18   7-segment incandescent displays. (Courtesy of Industrial Electronics Engineering, Inc.)

equipment with an expected life of 20,000 hours. Find the value of a dropping resistor to extend the life of the bulb to 20,000 hours. What is the intensity of the bulb in this condition?

FIGURE 3–19   Bulb circuit.

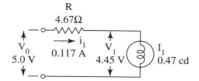

Solution

From Equation 3–9, $L_1 = L_0 \times (V_0/V_1)^{12}$, we can derive

$$V_1 = V_0/(L_1/L_0)^{1/12} = 5.00/(20,000/5,000)^{1/12} = 4.45 \text{ V}$$

At this voltage, the bulb current (Equation 3–10) is

$$i_1 = i_0\,(V_1/V_0)^{5.5} = 0.125 \times (4.45/5.00)^{5.5} = 0.117 \text{ A}$$

and the value of resistor R is

$$R = (V_0 - V_1)/i_1 = (5.00 - 4.45)/0.117 = 4.67 \text{ }\Omega$$

The intensity at reduced voltage is

$$I_1 = I_0(V_1/V_0)^{3.5} = 0.22 \times (4.45/5.00)^{3.5} = 0.147 \text{ cd}$$

Circuit is shown in Figure 3–19.

3–4 LIGHT EMITTING DIODE

The **light emitting diode (LED)** is definitely one of the most popular opto-electronics sources. It is inexpensive, consumes very little power, and is easily adaptable to today's electronics circuitry.

The operating principle of the LED is described on page 81. Actually, all semiconductor diodes produce radiation when electrons from the conduction band recombine with the holes in the valence band. In a normal silicon diode, the radiated wavelength is long (in the infrared range), and the radiation is absorbed by the surrounding semiconductor material. In an LED, the semiconductor has a high-energy gap and the junction is constructed so that radiation from the junction can escape.

The Construction of an LED

The typical construction of an LED is shown in Figure 3–20.

The LED is constructed on a GaP or GaAsP n-doped substrate. A thin epitaxial p-doped layer is grown on the top of this substrate. The p-n junction, where the recombination takes place and radiation is generated, is between these two layers. Since the GaP layer is transparent, the radiation escapes through the top layer. With GaAs substrate, the flux directed to the bottom is

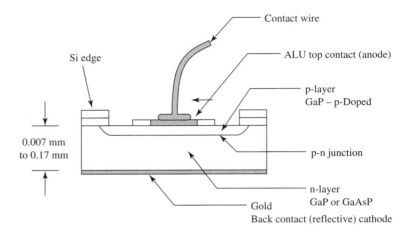

Si edge

0.007 mm
to 0.17 mm

Contact wire

ALU top contact (anode)

p-layer
GaP – p-Doped

p-n junction

n-layer
GaP or GaAsP

Gold

Back contact (reflective) cathode

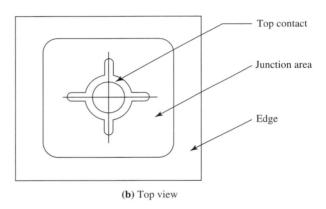

Top contact

Junction area

Edge

(b) Top view

FIGURE 3–20   Construction of an LED.

absorbed. With the GaP substrate that is transparent, a reflective layer is added to the bottom electrode to improve efficiency. The flux escape mechanism is shown in Figure 3–21(a).

The usable flux escapes from the top of the junction, through the p-layer. Applying Snell's law to the air and GaP interface we will find that, because of the complete internal reflection, only flux in a very narrow cone (with a cone angle of 17°) can escape, as seen in Figure 3–21(b). To improve this condition, the junction is encapsulated in plastic, increasing the escape angle to 26°. The encapsulation not only improves the optical efficiency, but also holds all the components of the LED together, as seen in Figure 3–22. The shape of the plastic dome, or lens, also controls the radiation pattern of the LED.

In Figure 3–22, some typical LED forms with their radiation patterns are shown. Figures 3–21(a) and (b) are called T-1 3/4 and T-1 envelopes,

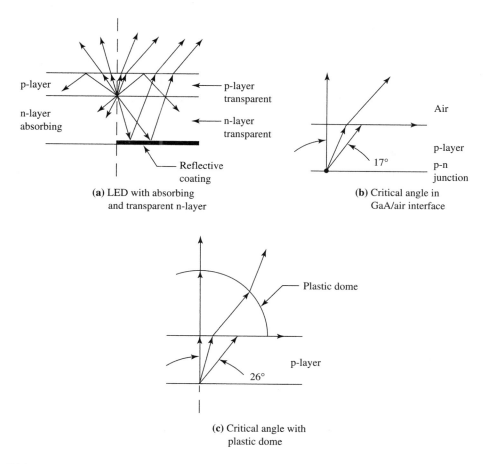

p-layer

n-layer
absorbing

p-layer
transparent

n-layer
transparent

Reflective
coating

(a) LED with absorbing
and transparent n-layer

Air

p-layer

p-n
junction

17°

(b) Critical angle in
GaA/air interface

Plastic dome

p-layer

26°

(c) Critical angle with
plastic dome

FIGURE 3–21   Optical design of an LED.

respectively, where the number after the T indicates the diameter of the LED in 1/8 of an inch.

In addition to the examples shown, LEDs are produced (because of their popularity) in many other configurations. They are also available with internal current-limiting resistors, two-color combinations, and many other variations.

Electrical and Optical Characteristics of an LED

The LED is a semiconductor diode. Therefore, its characteristics and limitations are similar to a normal *p-n* junction diode (depicted in Figure 3–23). Part (a) of the figure shows a typical diode voltage (V_D), forward current (i_D) characteristic.

The **breaking voltage** (V_C) is about 1.2 to 2 V, depending on the semiconductor material. The dynamic resistance ranges from a few ohms to tens

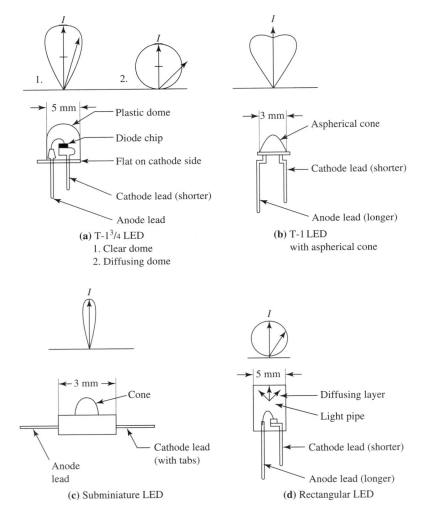

FIGURE 3–22   Examples of common LED lamps.

of ohms. The **reverse breakdown voltage** (V_B) is about 5 V. Figure 3–23(b) shows an important limiting factor, the **maximum power dissipation** (P_{MAX}) and its dependence on ambient temperature. The third parameter to consider, especially since the LED is often used in pulsed mode, is the **maximum permissible peak current** (i_{MP}). For pulsed operation, the relationship among these parameters is given in the following equations.

Duty cycle:

$$d_c = t_{on}/T \tag{3-12}$$

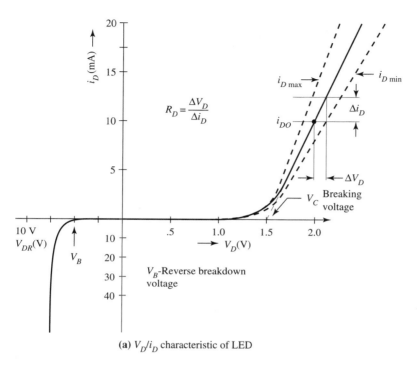

(a) V_D/i_D characteristic of LED

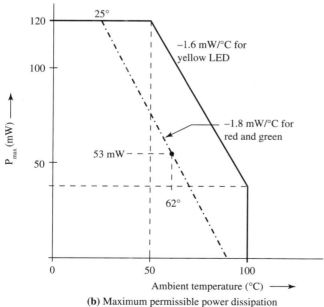

(b) Maximum permissible power dissipation

FIGURE 3–23   LED i_D/V_D and dissipation characteristics.

FIGURE 3-24   Duty cycle and peak and average current in a pulse train.

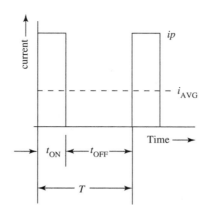

where d_c = duty cycle
$\quad t_{\text{on}}$ = pulse ON time (s)
$\quad T$ = pulse period (s).

The average current in the pulse train is

$$i_{\text{avg}} = i_P d_c \qquad\qquad\qquad (3\text{-}13)$$

where i_{avg} = average current of the pulse train (A)
$\quad i_P$ = peak current (A).

The LED power dissipation at steady state is

$$P_D = i_D V_D \qquad\qquad\qquad (3\text{-}14)$$

where P_D = steady state power dissipation (W)
$\quad i_D$ = LED current (A)
$\quad V_D$ = LED voltage (V).

For pulsed condition, the average power dissipation is

$$P_{\text{avg}} = i_{\text{avg}}[V_{DO} + R_D(i_P - i_{DO})] \qquad\qquad\qquad (3\text{-}15)$$

where P_{avg} = average power dissipation (W)
$\quad i_{\text{avg}}$ = average current (A)
$\quad V_{DO}$ = voltage at reference point (V)
$\quad i_{DO}$ = current at reference point (A)
$\quad R_D$ = LED dynamic resistance (Ω).

LED dynamic resistance can be calculated from the LED characteristic curve in Figure 3-23:

$$R_D = \Delta V_D / \Delta i_D \qquad\qquad\qquad (3\text{-}16)$$

where ΔV_D = voltage increment at the reference point (V)
$\quad \Delta i_D$ = current increment at the reference point (A).

Example 3-4 demonstrates the use of given relationships.

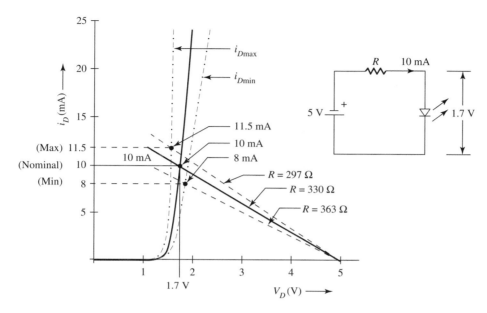

FIGURE 3–25　　Circuit of Example 3–4.

EXAMPLE 3–4

Calculate LED maximum and minimum current using 10% tolerance resistor in an LED with the characteristics shown in Figure 3–25.

Solution

The voltage drop across the resistor R is

$$V_\text{R} = V_\text{CC} - V_\text{D} = 5.00 - 1.70 = 3.30 \text{ V}$$

The nominal resistance value is

$$R = V_\text{R}/i_\text{D} = 3.30/0.010 = 330 \ \Omega$$

Thus $R_\text{max} = 3.30 \times 1.10 = 363 \ \Omega$ and $R_\text{min} = 3.30 \times 0.90 = 297 \ \Omega$.

Drawing the lines for maximum and minimum resistance values into the graph, we can find the crossing points of i_Dmin and i_Dmax curves, and maximum and minimum LED current:

$$i_\text{Dmax} = 11.5 \text{ mA} \quad \text{and} \quad i_\text{Dmin} = 8.0 \text{ mA}$$

Using components with normal tolerances, a variation of almost $\pm20\%$ in LED current can be expected.

The effect of an almost 20% current variation on LED brightness will be investigated later. That effect demonstrates the need for tightly controlled components when uniform brightness is required.

EXAMPLE 3–5

The LED in Example 3–4 is used in pulsed operation with 100 mA peak current and 10% duty cycle. Find the maximum permissible ambient temperature for a red LED.

Solution

From Equation 3–13, the average LED current is

$$i_{\text{avg}} = i_{\text{P}} d_{\text{C}} = 100 \times 0.1 = 10.0 \text{ mA}$$

From Figure 3–24 we find the dynamic resistance of the LED $R_{\text{D}} = 40 \ \Omega$. Using Equation 3–15, we can calculate the average dissipation thusly

$$P_{\text{avg}} = i_{\text{avg}}[V_{\text{DO}} + R_{\text{D}}(i_{\text{P}} - i_{\text{VO}})] = 0.010[1.70 + 40(0.100 - 0.010)] = 0.053 \text{ W}$$

Inserting this value into Figure 3–23(b), we find the maximum ambient temperature to be 62°C.

Intensity is the most important optical parameter of an LED. The intensity is a nonlinear function of LED current, so the relative intensity increases with increasing current. The nonlinearity can be expressed in two ways: by presenting an $I_{\text{PR}} = f(I_{\text{D}})$ (curve A, Figure 3–26) or by defining **relative efficiency** as $\eta_{\text{PR}} = I_{\text{PR}}/I_{\text{PRO}}$, as shown on curve B in Figure 3–25.

Curve B shows that at higher current the LED efficiency increases considerably. This fact supports pulsed operation of an LED since increasing the current in steady state condition is limited by the maximum power dissipation. When operating in pulsed condition, the average intensity can be calculated using the following equation:

$$I_{\text{PRavg}} = I_{\text{PRO}}(i_{\text{P}} d_{\text{C}} \eta_{\text{PR}}/i_{\text{O}} \eta_{\text{PRO}}) \tag{3-17}$$

where I_{PRavg} = average radiation intensity (cd, W/sr)
$\quad\quad I_{\text{PRO}}$ = reference radiation intensity (cd, W/sr)
$\quad\quad i_{\text{P}}$ = peak current (A)
$\quad\quad i_{\text{O}}$ = reference current (A)
$\quad\quad d_{\text{C}}$ = duty cycle
$\quad\quad \eta_{\text{PR}}$ = efficiency at peak current
$\quad\quad \eta_{\text{PRO}}$ = reference efficiency $\eta_{\text{PRO}} = 1$.

An application of this relationship is demonstrated in the next example.

EXAMPLE 3–6

The LED has a reference intensity at $i_{\text{O}} = 10$ mA of 12 mcd. The relative efficiency at 100 mA is $\eta = 1.5$. Find the LED average intensity for $d = 0.1$.

Solution

If LED is operated at reference condition, we have

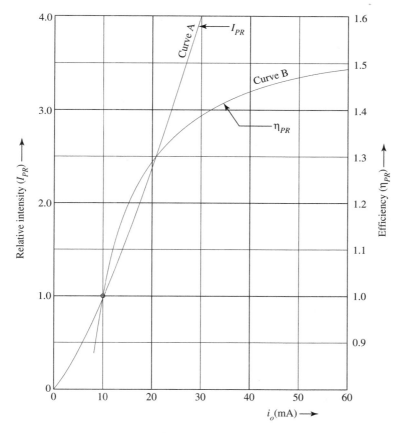

FIGURE 3–26   Relative LED luminous intensity and efficiency.

$$I_{\text{PRO}} = 12 \text{ mcd}; i_{\text{O}} = 10 \text{ mA}; \eta_{\text{PRO}} = 1$$

At peak current, we obtain

$$i_{\text{P}} = 100 \text{ mA}; d_{\text{C}} = 0.1; \eta_{\text{PR}} = 1.5$$

Using Equation 3–17, we get

$$I_{\text{PRavg}} = I_{\text{PRO}} \, (i_{\text{P}} d_{\text{C}} \eta_{\text{PR}} / i_{\text{O}} \eta_{\text{PRO}})$$
$$= 12(100 \times 0.1 \times 1.5)/10 \times 1) = 18 \text{ mcd}$$

As seen from the example, an LED operated at the same average current as the current at reference condition has 50% higher intensity when operated in a pulsed condition.

The typical **output spectra** of an LED is shown in Figure 3–26.

As we see from the curves, LED output is not monochromatic but contains considerable bandwidth around the central wavelength.

FIGURE 3–27   Typical
output spectra of an LED.

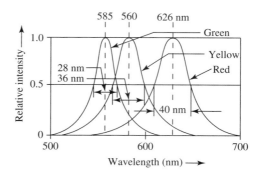

Another noteworthy characteristic of the LED is the temperature dependency of its luminous intensity. The **temperature coefficient** of I_{RP} is about $-1\%/°C$. A 25°C temperature rise will decrease the intensity by 25%. This factor cannot be overlooked.

The last, and quite favorable, characteristic of an LED is its fast **response time,** typically 90 ns for yellow and red and 500 ns for green. This short response time makes LEDs usable as sources in communications links. Incandescent sources, with very slow response time, have almost no application in this field.

LED Application Notes

The optoelectronics designer using LEDs must consider three design factors:

1. Proper mechanical configuration and radiation pattern
2. Dissipation and thermal condition
3. Drive circuit requirements

Factor 1 can easily be solved by consulting manufacturers' catalogs, where a great variety of LEDs for all possible configurations can be found. Factor 2, dissipation and thermal considerations, has already been discussed (see p. 100).

The purpose of the drive circuit (factor 3) is to guarantee desired luminous intensity from the LED. Intensity is a direct function of LED current so the design problem is to create a circuit that drives the desired current through the LED diode. Some typical and popular circuits are shown in Figure 3–28.

The simplest drive circuit is unquestionably the resistor-driving circuit, which uses only one resistor. The disadvantage of this circuit is that, due to the tolerances in LED characteristics, resistance value, and variation of supply voltage, the LED current and resulting intensity may exhibit considerable variations. This factor may be of no consequence when the LED is used as a single indicator. However, in an array where several LEDs are situated

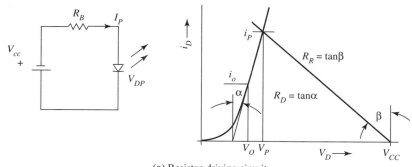

(a) Resistor-driving circuit

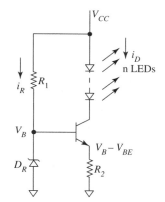

(b) Transistor constant current drive

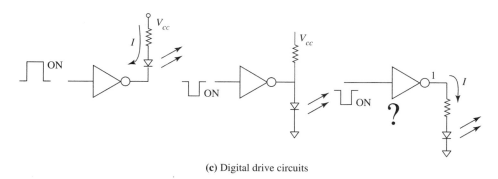

(c) Digital drive circuits

FIGURE 3–28   LED drive circuits.

next to each other, the intensity variations may be objectionable. The value of the dropping resistor R_B may be calculated using the following equation:

$$R_\text{B} = (V_\text{CC} - V_\text{O})/i_\text{P} - R_\text{D}(i_\text{P} - i_\text{O})/i_\text{P} \qquad \textbf{(3–18)}$$

The quantities in the equation are explained in Figure 3–28. Several LEDs may be connected in series using this simple circuit. Parallel connection of LEDs, however, should be avoided since the variations in LED characteristics cause unacceptably dangerous variations in LED current.

EXAMPLE 3–7

An LED is driven by a $V_{cc} = 12$ V source. The current should be $i_p = 30$ mA. At $i_o = 10$ mA, $V_o = 2.2$ V and the LED dynamic resistance is $R_D = 15$ Ω. Find the driving resistance R_B.

Solution
Using Equation 3–18, we derive

$$R_B = (V_{cc} - V_o)/i_P - R_D(i_p - i_o)/i_p$$
$$= (12.0 - 2.20)/30.0 \times 10^{-3} - 15.0(30.0 - 10.0)/30.0 = 316 \ \Omega$$

The reference current, voltage, and dynamic resistance of an LED are normally not given on the data sheet, so they must be derived from the diode curve. Therefore, for all practical purposes, using the graphic solution shown in Figure 3–28(a) is simpler and faster.

With a few more components and a transistor, we can design a constant current drive that guarantees a predictable and stable current through a single LED or an array. The advantage of this circuit is that the current is not at the mercy of LED characteristic variations. The circuit is shown in Figure 3–28(b). It can be used for many LEDs in series, as long as the total voltage drop across the LEDs does not exceed the available collector voltage V_{CC}. The circuit can be designed using the following relationships:

$$R_1 = (V_{CC} - V_B)i_R \tag{3–19}$$

where V_{CC} = supply voltage (V)
$\qquad V_B$ = voltage of the reference diode (V)
$\qquad i_B$ = current through the reference diode (A).

$$R_2 = (V_B - V_{BE})/i_D \tag{3–20}$$

where V_{BE} = base-emitter voltage drop $V_{BE} = 0.7$ (V)
$\qquad i_D$ = desired LED current (A).

For the circuit to operate properly, the following condition must be met:

$$V_{CC} > nV_P + (V_B - V_{BE}) + 0.2 \text{ V}$$

EXAMPLE 3–8

Five LEDs with $i_D = 30$ mA are driven from a 24 V source. Design the constant current-drive circuit.

Solution

Several components may vary in this design: for example, the zener diode voltage and current. Higher voltage reduces the temperature effect of the emitter junction, but in most cases lower voltage of few volts is satisfactory. Sometimes even a GaAsP (red) LED with 1.5 V voltage drop is very satisfactory since it compensates for the temperature coefficient of the emitter junction. A 1 mA current is sufficient to keep a zener diode in stable condition. Selecting $V_B = 2.2$ V and using Equation 3–19, we get

$$R_1 = (V_{cc} - V_B)/1 \text{ mA} = (25.0 - 2.2)/1 \times 10^{-3} = 22.8 \text{ k}\Omega$$

The emitter voltage is

$$V_E = V_B - V_{BE} = 2.2 - 0.7 = 1.5 \text{ V}$$

The value of R_2 using Equation 3–20 is

$$R_2 = V_{BE}/I_D = 1.5/30 \times 10^{-3} = 50 \ \Omega$$

Check that total voltage drop across the LEDs does not exceed available V_{cc} − collector voltage, which is

$$V_{cc} - V_c = 24.0 - 1.5 - 0.2 = 22.3 \text{ V}$$

Five LEDs will require $5 \times 1.5 = 7.50$ V, which is less than 22.3 V.

Figure 3–28(c) shows several driving circuits using digital logic. This application is very common. All logic families have limited HIGH output current capabilities. Therefore, driving an LED from active HIGH current is not possible. The sinking current is somewhat higher (16 mA for TTL), and therefore the LED can be driven with modest current using active LOW, as shown in the first two circuits in Figure 3–28(c). When higher current is required, as in pulsed and multiplexing modes, for example, special drivers must be used.

3–5 ELECTROLUMINESCENT SOURCES

There are two types of electroluminescent sources: the **injection** or **DC source** (the LED), and the **intrinsic** or **AC source** (the electroluminescent panel). Both work on the same principle—excitation of electrons to a higher energy level by electric field.

The electroluminescent panel essentially is a capacitor where the dielectric is dispersed with an **electroluminescent pigment,** usually a semiconducting phosphor. When AC is applied to the electrodes, radiation is emitted. To allow the radiation to escape, one of the electrodes of the capacitor is made translucent.

This simple technology is used in electroluminescent lighting and display panels, described in the following sections.

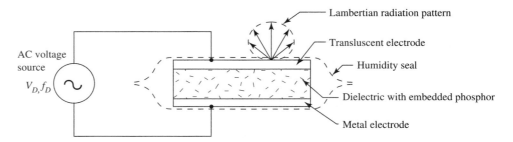

FIGURE 3–29   Electroluminescent lighting panel.

Electroluminescent Lighting Panel

As mentioned before, the **electroluminescent panel** is basically a capacitor, as shown in Figure 3–29.

The device is sealed, to protect the dielectric from humidity in the air.

The panel has many uses, such as back lighting for LCD displays and instrument panels, lighted instrument panels, warning and exit signs, and so on. Panels come in voltage ratings from 20 to 400 V and for operation frequency from 30 to 400 Hz. The source luminance is a function of voltage, frequency, and the color of the source. A typical relationship is shown in Figure 3–30.

As seen from the curves, the luminescence (or brightness) of the panel greatly depends upon driving voltage and frequency. The luminance also is the function of temperature, as shown in Figure 3–31.

The panels do have an undesirable characteristic: their luminance greatly decreases with age. A typical aging curve is shown in Figure 3–32. The rapidly deteriorating brightness can be somewhat compensated for by

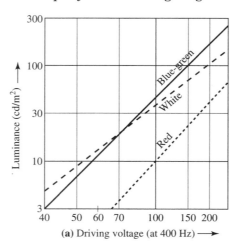

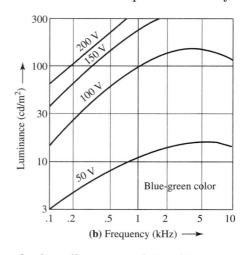

FIGURE 3–30   Lighting panel luminance and voltage/frequency relationship.

initially driving the panel with less than maximum luminance and by adding brightness control to the circuit, so the driving voltage may be increased as the panel ages.

Typical electrical characteristics of a 115 V rated panel are shown in Table 3–3.

As one can see from Table 3–3, the panel requires medium power but, as mentioned before, high voltage, too. The converters to change DC to proper driving voltage and frequency are available.

The panel produces single-wavelength (or, in the case of white or less saturated blue) double-wavelength spectra. Typical radiation characteristics are shown in Table 3–4.

FIGURE 3–31 Lighting panel luminance and temperature relationship.

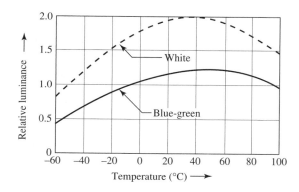

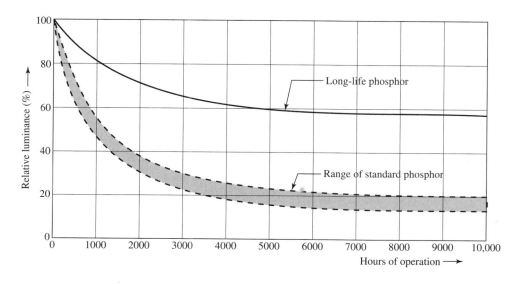

FIGURE 3–32 Change of luminance during the life of electroluminescent panel (at 115 V, 400 Hz).

TABLE 3–3
Electrical Characteristics of a Luminescent Panel

Parameter	Value	Unit
Capacitance	4.1	nF/in^2
Phase angle	80	degree lead
Operating frequency	60	400 Hz
Nominal current	0.3–1.7	mA/in^2
Volt amperes	0.035–0.195	VA/in^2
Power	5–35	mW/in^2

TABLE 3–4
Spectral Output from a Luminescent Panel

Hue	Driving Frequency Hz	Luminance cd/m^2	Peak Emission nm
Blue-green	60	14	515
Blue-green	400	68	505
Aviation green	60	14	530
Aviation green	400	68	520
Yellow green	400	25	540
Yellow orange	400	16	580
Blue	400	13	470
White	400	50	NA
Blue white	400	41	NA
Red	400	14	620

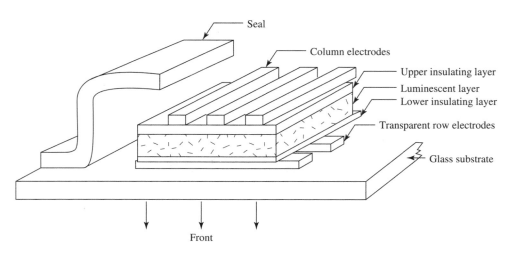

FIGURE 3–33   Electroluminescent dot matrix display.

The locations of panel radiation patterns are plotted in Figure 3–12, the chromaticity diagram.

Electroluminescent Display

The panel design, with slight modification, can be successfully employed in a **dot matrix panel,** as shown in Figure 3–33.

When appropriate voltage is applied to row and column electrodes, the luminescent layer between the electrodes glows, forming a **display cell,** called a **pixel,** ranging in size from approximately 0.02 by 0.03 inches to as small as 0.01 by 0.01 inches. The display color is amber, 584 nm, and the intensity is about 75 cd/m^2. As with electroluminescent lamps, brightness deteriorates with time.

Displays come in a variety of sizes from 65,000 to 250,000 pixel elements. With a high number of pixels high resolution (3 lines/mm) can be achieved. This level of readability competes with a cathode ray tube display. While several colors (orange, green, red, and yellow) are available, the orange color has the highest luminance.

The display requires a high voltage, pulsed-power supply and a microprocessor-based control circuit. Those circuits are supplied with the display panel by the manufacturer. A typical panel is shown in Figure 3–34.

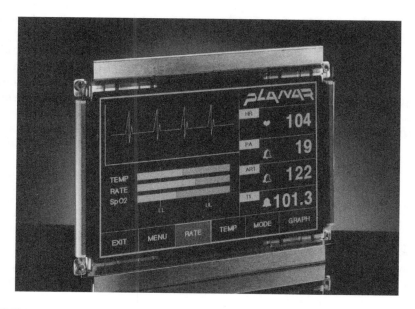

FIGURE 3–34   Typical electroluminescent display panel. (Courtesy of Planar Systems, Inc.)

3–6   VACUUM FLUORESCENT SOURCE

In a **vacuum fluorescent (VF) source,** the radiation originates from a semiconductor, usually a phosphor, that is bombarded by high-speed electrons in a vacuum. As a result, the phosphor's valence electrons are elevated to a higher energy level. When the electrons fall back to the ground energy state, a narrow band of visible radiation is emitted.

Operating Principles

The VF source is very similar to a vacuum tube, as seen in Figure 3–35.

The electrons from a heated cathode are accelerated through a grid to an anode where they strike light-emitting material. The electron flow is controlled by the grid potential: a positive grid passes and accelerates the electrons; a negative grid blocks the electron flow. Sometimes the grid and anode are connected.

This technology is applied in two ways. In VF displays, the viewer sees the display through the grid, from the filament side. In the cathode ray tube version, the viewer sees the display from the side of the emitting phosphor.

Vacuum Fluorescent Display

The VF display is available in segmented and dot matrix versions, bar graph displays, or X-Y grid form for graphic displays. The construction of these displays differs, as shown in Figure 3–36.

In a segmented display [Figure 3–36(a)], all digits can be individually turned on or off by applying a positive or negative potential to the grid, allowing only selected digits to be displayed. Segments of the display are energized by applying positive potential. Segments of all digits may be connected in parallel, allowing simple multiplexing of the digits.

In the dot matrix display [Figure 3–36(b)], the arrangement is similar, except for an extra grid raster added between the digit and anode grid. This, together with the anode raster, forms the dot matrix.

Figure 3–36(c) shows an X-Y display. The column grid raster and the row anode raster allow any single pixel to be energized, forming a graphic grid.

These displays, because of their modest cost, are very popular for clocks, desk calculators, annunciators, etc. They come either with or without driving circuitry. The driving methods are described in Chapter 5.

To operate the display, a low voltage AC filament supply (below 10 V) and less than a few hundred milliamps are needed. The anode drive is a positive supply of a few tens of volts that requires tens of milliamps. The grid voltage is of the same magnitude.

The luminance of the display is up to 1000 cd/m^2, depending on the anode voltage. A typical anode voltage-luminance relationship is shown in Figure 3–37.

The color of the display depends upon the phosphors used. Some available colors are shown in Figure 3–38. The emission is relatively wide band, single frequency. The chroma of the examples is shown on the C.I.E. chromatic diagram on page 86.

The filament is driven at low temperature, about 700°C, below the glowing point of the cathode. Therefore, the life of the device is long, measured in tens of thousands of hours. There is very little deterioration of the intensity with time. Filters can be used to modify the color of the display and reduce glare. Figure 3–39 shows a typical VF display.

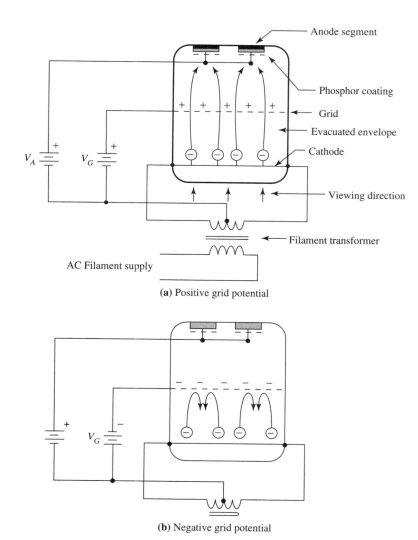

FIGURE 3–35   Vacuum fluorescent display.

The Cathode Ray Tube

The **cathode ray tube (CRT)** is the most versatile and highest resolution display device. When a display of more than a few thousand characters is required, the CRT is unsurpassed. The CRT's disadvantage is its large size and high voltage requirement.

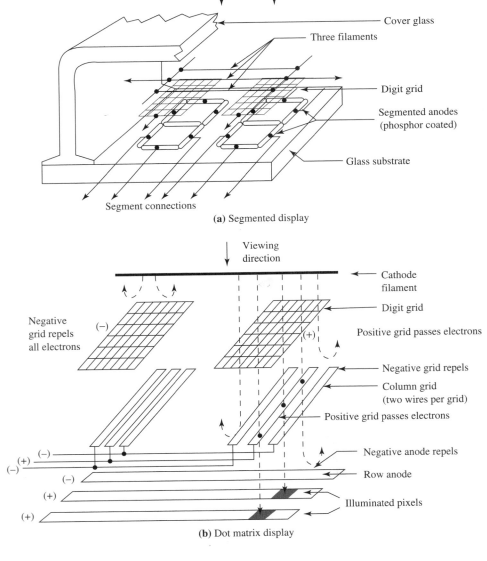

(a) Segmented display

(b) Dot matrix display

FIGURE 3–36a, b   Segmented and dot matrix VF displays.

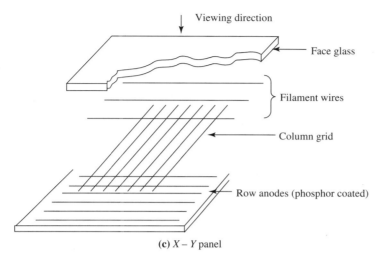

(c) *X – Y* panel

FIGURE 3–36c   *(Continued)* Segmented and dot matrix VF displays.

The CRT is an evacuated glass envelope, as shown in Figure 3–40. At one end is an **electron gun**—a heated cathode that emits electrons. The electrons are accelerated toward a positively charged anode. A negatively biased screen controls the electron flow. The electrons pass through a complex **focusing system** [see Figure 3–40(b)] that concentrates them into a narrow beam focused on the screen at the other end of the envelope. The

FIGURE 3–37   VF display anode voltage/luminance relationship.

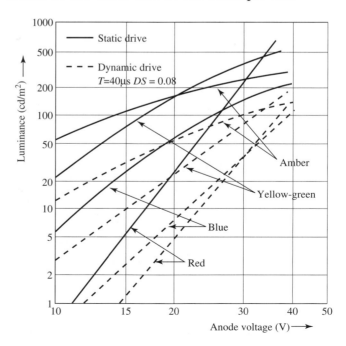

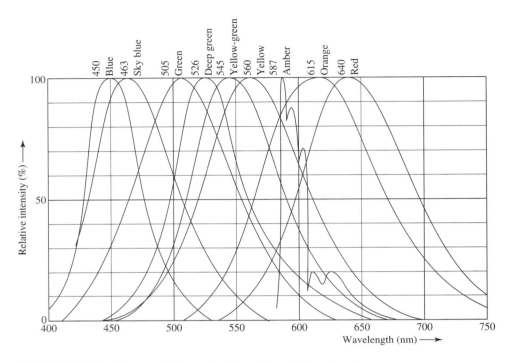

FIGURE 3-38   Spectral characteristics of available phosphors.

inside face of the screen is coated with phosphor that emits light upon the impact of the high-speed electrons, thus producing a visible spot.

The electron beam, after emerging from the gun, is passed between two pairs of **deflection plates,** or a **magnetic deflection device.** The plates deflect the beam so that the beam makes a tightly spaced horizontal line pattern on the screen. The horizontal lines can be converted into information pixels by modulating the beam intensity using the control grid, forming a tightly spaced pixel pattern.

FIGURE 3-39   Typical VF display. (Courtesy Industrial Electronics Engineers, Inc.)

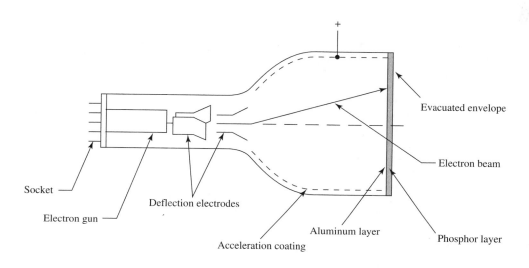

(a) Parts of the cathode ray tube

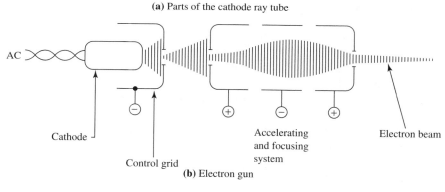

(b) Electron gun

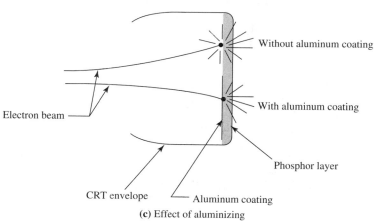

(c) Effect of aluminizing

FIGURE 3–40   The cathode ray tube.

TABLE 3–5
Common CRT Phosphor Characteristics

Phosphor Number	Color	Relative Visual	Speed Photo	Persistence	Typical Application
P1	Yellow-green	45	35	Medium	Scopes, radar
P4	White	50	75	Medium	Monochromatic TV
P31	Green	100	75	Medium-short	Scopes
P33	Orange	20	7	Long	Radar
P35	Blue-white	55	45	Medium-short	Scopes
P37	Green-blue	—	—	Very short	Scanners
P45	White	—	—	Medium	High brightness displays

Two types of phosphor screens are available: plain and aluminized screen, as seen in Figure 3–40(c). In the plain screen, the light from the bombarded spot radiates in two directions and half of the radiation is lost. With the aluminized screen, electrons pass through a very thin aluminum coating inside the screen with some loss of energy, but all the radiation is reflected toward the front face of the screen, which results in higher spot intensity. Aluminizing also protects the screen from burnout, which is a common problem when a high-energy spot is left stationary.

The color of the display depends upon the phosphor used. Over fifty types of phosphors are available for CRT use. Common phosphors and their characteristics for monochromatic displays are listed in Table 3–5.

The phosphor emits a single-frequency, wide-bandwidth spectrum. Figure 3–41 depicts typical radiant energy distribution of some phosphors.

In all CRT displays, the electron beam is scanned across the screen, so the radiation from the phosphor can be divided into two periods, as seen in Figure 3–42.

First is the period during which the phosphor is excited with the electron beam. During this period, the phosphor exhibits **fluorescence.**

FIGURE 3–41   Typical phosphors spectra.

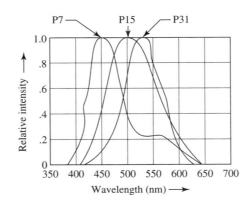

FIGURE 3–42 Fluorescent and phosphorescent phase in phosphor.

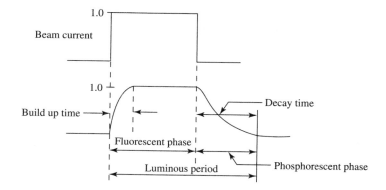

The electron-beam energy is converted directly into luminous flux. In the following period, after the excitation is finished, most phosphors still exhibit afterglow, a process called **phosphorescence.**

The decay time of the phosphorescence depends upon the type of phosphors used. It may extend to several seconds, allowing the construction of a **long persistence screen,** where the image is held for a long time. The decay characteristic of type P31, the most common phosphor for oscilloscopes, is shown in Figure 3–43.

For some phosphors, the fluorescent and phosphorescent colors are different. For P32 type phosphor, the fluorescent radiation is purplish blue and the phosphorescence is yellowish green.

The intensity of the spot depends upon electron beam current density and acceleration voltage. This relationship is depicted in Figure 3–44.

FIGURE 3–43
Phosphorescent decay of P31-type phosphor.

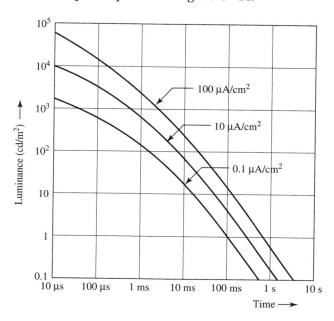

FIGURE 3–44 Luminance characteristics of various phosphors.

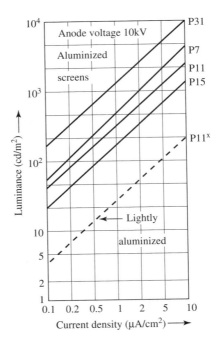

Table 3–5 lists relative intensities, with phosphor P31 having a reference intensity of 100. Since the CRT tube image is often photographed, relative photographic speed is also given. Because human vision response is different from the photographic material, the intensity and speed figures differ.

The location of listed phosphor colors is shown on the C.I.E. chromatic diagram on page 86.

Color display requires a different CRT tube. The most common color tube has three electron guns, as seen in Figure 3–45.

The beams from the three guns are directed through a hole in the **shadow mask** in front of the screen. Thus, three dots are formed on the screen. The screen is covered with a triangular phosphor dot pattern. Each dot in the pattern emits a primary color: blue, green, or red. The screen pattern and the shadow mask are precisely aligned so that one gun always excites one color. Therefore the intensity of each primary color can be controlled. By combining the three beams at different intensities, all chromas can be created. A typical color CRT tube screen has about 200,000 three-dot patterns.

The location of the primary colors, blue, red, and green (marked B′, R′, and G′) is shown on the C.I.E. chromatic diagram on page 86. As seen, the primaries form a triangle within the diagram. By varying the intensity of the primary colors, all hues and saturations within this triangle can be reproduced.

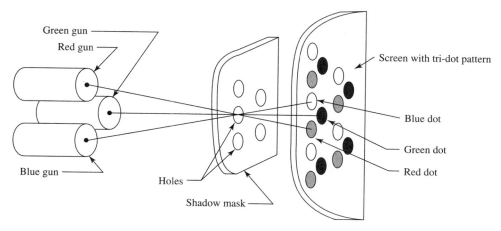

FIGURE 3–45   Principle of color CRT tube.

3-7 GAS-DISCHARGE LAMPS

Gas-discharge lamps, once a mainstay of numerical display under the Nixie trade name, have lost their importance because their bulk and high-voltage requirements are not compatible with modern electronics circuitry. Nevertheless, the lamps are still used as AC-powered pilot lights and in some panel displays.

Operating Principle and Electrical Characteristics

The **gas-discharge lamp** is a glass envelope filled with low-pressure gas, typically neon, and two electrodes, as shown in Figure 3–46.

When a low-voltage source is connected to the electrodes, the lamp acts as an insulator. However, when the voltage is increased (usually over 90 V),

FIGURE 3–46 Gas discharge lamp and its potential distribution.

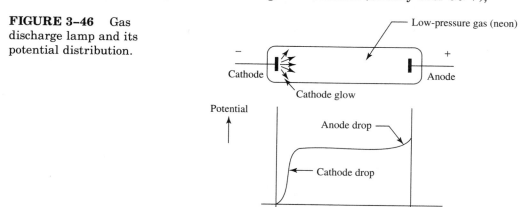

the gas is ionized and conducts electricity. The potential distribution along the lamp in this condition is shown in Figure 3–46. As seen from the graph, a sharp potential gradient develops close to the cathode. This accelerates electrons emitted from the cathode and brings the gas near the cathode to an excited state. As a result, a glow develops around the cathode electrode.

The electrical characteristics of the gas-discharge lamp are shown in Figure 3–47.

As seen from the graph, the lamp acts as an insulator until the applied voltage reaches the ignition voltage (about 80 to 90 V). At this point, the lamp ignites and the voltage drops to operating level (about 70 V). Therefore, the lamp is a nonlinear resistor and always needs a limiting resistor to operate. The value of the resistor for given lamp current can be calculated from the equation

$$R_L = (V_S - V_O)/i_L \qquad\qquad (3–21)$$

where R_L = limiting resistor (Ω)
     V_S = source voltage (V)
     V_O = lamp operating voltage (V)
     i_L = lamp current (A).

When the lamp is constructed symmetrically (that is, the cathode and anode electrodes are alike, as in the case of a pilot lamp), the lamp can also be driven with AC. In this case, both electrodes will light up.

The color of the lamp depends upon the gas filling. With neon, the most common gas, it is yellow. With helium, the color is red. Since the radiation contains ultraviolet wavelengths, other colors such as blue, green, and orange can be created by coating the inside of the lamp with phosphor.

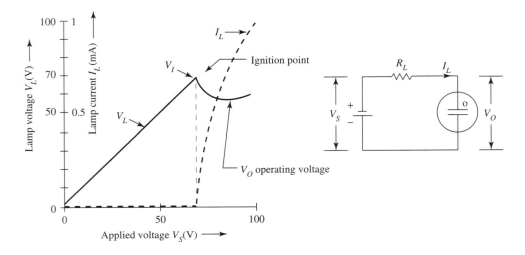

FIGURE 3–47   Electrical characteristics of gas discharge lamp.

FIGURE 3–48 AC or DC plasma display.

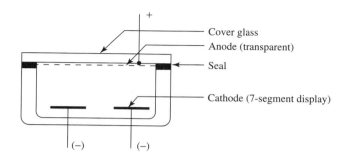

Cover glass
Anode (transparent)
Seal
Cathode (7-segment display)

Plasma Display

The modern version of the same lamp is designed in a flat planar form as a segmented display, as seen in Figure 3–48.

A slightly different principle is used in the **AC plasma discharge display,** as shown in Figure 3–49.

In the AC plasma display the electrodes are shielded with a dielectric layer that has memory, leading to improved brightness. This system is also applicable to flat panel, segmented, and dot matrix displays.

Plasma displays, with some refinements, are resulting in the development of larger dot matrix displays. Their general advantage is that luminance reaches 500 cd/m^2, far superior to other techniques. Also, they can be produced with fairly high resolution, up to one million pixels. They are excellent for large displays viewed at distances of up to 50 feet.

Plasma devices require high voltage, typically 200 V DC. This problem is solved by DC to DC converters that are usually built into the driver circuit boards. Also, separate compact converters especially designed for this purpose are available. The current requirement is several milliamps per digit. Because of the high voltage, special driver gates are needed. A typical plasma is shown in Figure 3–50.

FIGURE 3–49 Flat 7-segment plasma display.

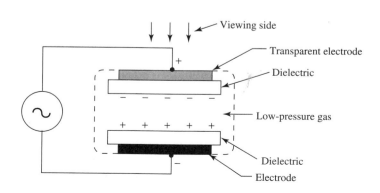

Viewing side
Transparent electrode
Dielectric
Low-pressure gas
Dielectric
Electrode

FIGURE 3–50   Typical plasma display panel. (Courtesy Industrial Electronics Engineers, Inc.)

SUMMARY

Chapter 3 explains how radiation is created, emphasizing the semiconductor junction. A set of definitions is introduced to enable the user to classify the sources by flux output, wavelength and color, radiation spectrum, and mode of excitation. The C.I.E. chromatic diagram was introduced and the term **color temperature** explained. A table was compiled that compares the pertinent characteristics of the most common radiation sources.

Further, six sources — incandescent lamps, LEDs, electroluminescent sources, vacuum fluorescent sources, cathode ray tubes, and gas discharge sources — are described, including their operating principles and technical characteristics. Detailed design principles and data are given for incandescent lamps and LEDs.

RECOMMENDED READING

Hewlett-Packard. 1981. *Optoelectronics Fiber Optics Application Manual.* New York: McGraw-Hill Book Company.

Bylander, E.G. 1979. *Electronic Displays.* New York: McGraw-Hill Book Company.

Kingslake, R. 1965. *Applied Optics and Optical Engineering,* Vol. 1. New York: Academic Press.

Driscoll, C., Editor. 1978. *Handbook of Optics.* New York: McGraw-Hill Book Company.

PROBLEMS

1. Calculate radiated wavelength when an electron in a hydrogen atom decays from the fourth orbit to the second orbit.

2. What wavelength is radiated from a silicon ($\Delta E = 1.09$ eV) and from a germanium ($\Delta E = 0.66$ eV) junction?

3. Describe the chroma (hue and saturation) of sources which have the following C.I.E. chromatic coordinates:

   a. $x = 3.10$, $y = 3.10$

   b. $x = 4.50$, $y = 1.30$

   c. $x = 2.00$, $y = 6.00$

4. A GE 100 W soft white light bulb delivers 1710 lm flux, has an average life of 750 h, and is designed to operate at 120 V.

   a. Find the bulb's efficacy and color temperature considering that 90% of bulb energy consumption is radiated.

   b. Calculate the value of a dropping resistor to extend the bulb's life to 2000 h and find color temperature and efficacy at this condition.

   c. Considering energy cost of 0.10 \$/kWh and bulb cost of \$0.50, is there any saving of illumination cost (lm/h) using a bulb with extended life?

5. A power transistor is used to switch a 5 W, 12 V bulb. What is the minimum peak current rating for the transistor?

6. An LED with $V_{D0} = 1.50$ V and $R_D = 50$ Ω is driven in pulsed circuit with 20% duty cycle and 100 mA peak current. Find average dissipation and highest permissible ambient temperature using Figure 3–23(b) dissipation curve for a green LED.

7. An LED intensity is 15 mcd at 10 mA current. The LED is driven with 50 mA peak current and 25% duty cycle. Using Figure 3–25 curves, find

   a. LED average intensity at pulsed condition

   b. LED intensity when driven with steady-state current that equals LED average current at pulsed condition.

8. Design a transistor drive circuit for 4 LED [as on Figure 3–37(a)]. Circuit conditions: $V_{cc} = 12$ V, $V_B = 2.7$ V, $V_D = 1.5$ V. Find the resistor R_2 value and check if V_{cc} is sufficient to drive 4 LEDs.

9. A blue/green color back-light panel with 15 cd/m^2 luminance is needed. A 200 Hz power supply is available. Find the driving voltage.

10. A blue vacuum fluorescent display having luminance of 20 cd/m^2 is needed. Find the anode voltage of the display when

    a. static drive is used

    b. $d_s = 0.08$ dynamic drive is used.

11. What is the CRT beam current that uses P15 (green) phosphorus, has spot diameter of 0.5 mm, and spot illumination of 300 cd/m²?

12. A gas discharge lamp with operating voltage of 75 V is driven from a 100 V source. Find the value for limiting resistor for 25 mA lamp current.

4
Lasers

Laser is an acronym for **Light Amplification by Stimulated Emission Radiation.** The acronym describes the operation of the technology. The laser is a unique source that simultaneously produces both coherent (in-phase) and monochromatic (single-wavelength) radiation. Laser light has propagation characteristics that make possible numerous applications that cannot be achieved with random or collimated sources.

Each light source has particular characteristics. Those characteristics are most easily seen by passing the light from different sources through the same lens. Figure 4–1 shows this process, using three different light sources.

Figure 4–1(a) depicts the image of a 1-mm circular source of incoherent light. For given focal length and object distance, the image is a circle of 0.33 mm in diameter.

Figure 4–1(b) depicts the image from a collimated source with a beam coming through a 1-mm diameter opening (light from a distant star, for example). Assuming a perfect lens, the beam should concentrate to a point, at the focal point of the lens — this, after all, is the definition of the focal point. Actually, due to the wave nature of the light, the beam does not form a perfect point but a diffraction circle with a diameter of 61 μm. Diffraction therefore limits the resolution of the lens or the concentration of the light energy.

Figure 4–1(c) depicts the same experiment performed with a laser (coherent light) source 10 m away. The image circle is now 5 μm in diameter, considerably smaller than that of a collimated source.

This unique characteristic gives the laser many new applications: for fine definition information transfer as in compact disk (CD) players and laser printers, high energy concentration applications in medical surgery, metal cutting, and military applications, to name a few. Additionally, the laser

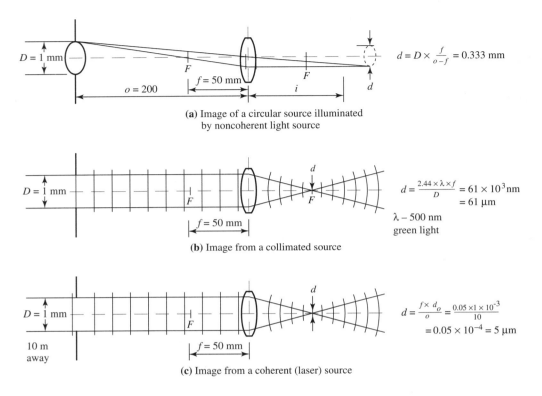

$$d = D \times \frac{f}{o-f} = 0.333 \text{ mm}$$

(a) Image of a circular source illuminated by noncoherent light source

$$d = \frac{2.44 \times \lambda \times f}{D} = 61 \times 10^3 \text{ nm}$$
$$= 61 \text{ μm}$$

$\lambda - 500$ nm
green light

(b) Image from a collimated source

$$d = \frac{f \times d_o}{o} = \frac{0.05 \times 1 \times 10^{-3}}{10}$$
$$= 0.05 \times 10^{-4} = 5 \text{ μm}$$

(c) Image from a coherent (laser) source

FIGURE 4–1   Image forming from different light sources.

beam can travel long distances with minimum dispersion. This characteristic has many communications and metrology applications. The ultimate demonstration of this characteristic was the laser beam that was directed to the moon and bounced back, to measure the distance from the earth to the moon.

Another important property of the laser is the monochromatic nature of its radiation. This aspect opens many applications in the communications field, such as very wide bandwidth high frequency modulation and fiber-optics transmission. Without laser sources, many recent advances in these fields would not have been possible.

Laser theory and its applications in technology are both extensive and complex. In this chapter only the fundamentals are introduced: elements of laser optics, principles of laser operation, and types of lasers and their applications. Somewhat more attention is given to the laser diode because its use and applications are well within the discipline of electronics, and it is often used by electronics engineers.

4–1   ELEMENTS OF LASER OPTICS

The purpose of geometric optics is to describe object and image relationships in an optical system using lenses or mirrors. Laser beams seldom produce an image, so laser optics considers the laser beam itself, its cross section, and its energy concentration, or illuminance. All applications of lasers require controlling the laser beam, so understanding laser optics is important.

We will deal only with the simplest TEM_{00} mode of beam where the irradiance across the cross section of the beam follows Gaussian distribution:

$$E_r = E_0 e^{-2r^2}/r_0{}^2 \qquad (4\text{–}1)$$

where E_0 = irradiance at the center of the beam (W/m², lm/m²)
  E_r = irradiance at the distance r from the center (W/m², lm/m²)
  r_0 = distance to the beam edge (m).

The Gaussian beam is depicted in Figure 4–2.

The **beam edge** is where the irradiance falls to $1/e^2$, or 0.135 of the center value. The **beam width** (d_e) is

$$d_e = 2r_e \qquad (4\text{–}2)$$

Figure 4–1 shows a parallel or collimated beam. In practice this type of beam rarely exists, as shown later. A normal beam either **diverges** or **converges,** as shown in Figure 4–3.

The smallest diameter of a converging/diverging beam is called the **beam waist.** This region of the beam is important because, in many cases, the task of the designer is to concentrate the beam to as small a diameter as possible. The waist diameter depends on the angle of divergence and the wavelength of the beam:

$$d_0 = 4\lambda/\pi\Theta \qquad (4\text{–}3)$$

where d_0 = waist diameter (m)
  λ = wavelength of the radiation (m)
  Θ = divergence angle (rad).

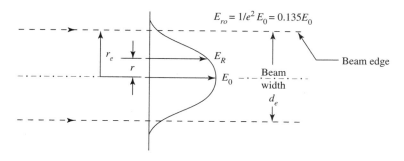

FIGURE 4–2   Cross section of TEM_{00} mode Gaussian beam.mmm

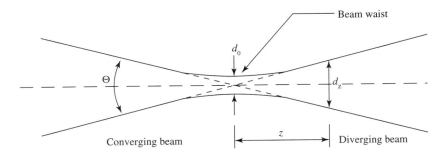

FIGURE 4-3   Diverging beam and beam waist.

The equation shows that the product of the waist diameter and the divergence angle is constant for a fixed wavelength:

$$d_0 \Theta = 4\lambda/\pi \qquad (4\text{-}4)$$

Thus, for a small waist diameter the divergence must be large. In the vicinity of the waist, the beam diameter is

$$d_z{}^2 = d_0{}^2 + \Theta^2 z^2 \qquad (4\text{-}5)$$

where d_z = waist diameter at the distance z (m)
   z = distance from the waist (m).

Farther away from the waist, where $\theta z >> d_0$, the equation simplifies to

$$d_z = \Theta z \qquad (4\text{-}6)$$

EXAMPLE 4-1

A hypothetical coherent source emits 500 nm of radiation at the divergence angle of 2°. Find the waist diameter and the beam diameter at 10 mm from the waist.

Solution
The divergence angle is $\Theta = 2° = 2/57.3 = 0.0349$ rad
From Equation 4-3: $d_0 = 4\lambda/\pi\theta = 4 \times 500 \times 10^{-9}/3.14 \times 0.0349$
$$= 18,240 \times 10^{-9} = 18.24 \ \mu m$$
At 10 mm from the waist:

$$d_{10} = \sqrt{(18.24 \times 10^{-6})^2 + (0.01 \times 0.0349)^2}$$
$$= \sqrt{332.7 \times 10^{-12} + 0.349^2 \times 10^{-6}} \approx 0.349 \times 10^{-3} \ m$$
$$= 0.349 \ (mm)$$

Often the laser beam is utilized in its waist region where the power concentration is highest. Therefore this region, also called the **Rayleigh range,** requires special attention. The Rayleigh range is the distance from the waist to where the beam diameter increases to $\sqrt{2d_0}$. Rearranging Equation 4–5, we can write for the beam diameter:

$$d_z = d_0\sqrt{1 + (\Theta \cdot z/d_0)^2} \tag{4-7}$$

Obviously $d_z = \sqrt{2d_0}$ when the second term under the radical equals 1. Therefore, the Rayleigh range is

$$z = z_R = d_0/\Theta \tag{4-8}$$

where z_R = Rayleigh range of the waist (m).

The condition of the beam within the Rayleigh range is shown in Figure 4–4.

Rayleigh range can also be described in the terms of wavefront curvature. At the waist the beam is parallel, and therefore the radius of the wavefront is infinity. On both sides of the Rayleigh range the beam looks as if it is coming from a point source, so the radius of curvature increases with distance from the waist. At the end of the Rayleigh range, the curvature has minimum radius. The **collimated region** of the beam can be taken from $-z_R$ to $+z_R$, or total distance of $2z_R$. Using previously derived equations, the Rayleigh range can also be expressed as

$$z_R = d_0/\Theta = 4\lambda/\pi\Theta^2 = \pi d_0{}^2/4\lambda \tag{4-9}$$

From the given equations, it is clear that whenever any two of the parameters z_R, θ, d_0, or λ are given, all other parameters can be derived, and the beam completely described. Within the Rayleigh range, the beam is considered to be collimated.

EXAMPLE 4–2

A typical He-Ne laser has beam waist of $d_0 = 1.1$ mm and radiates at 632.8 nm. Find the divergence angle and Rayleigh range.

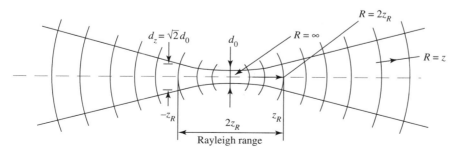

FIGURE 4–4   Rayleigh range.

Solution

From Equation 4–4:

$$\theta = 4\lambda/\pi d_0 = 4 \times 632.8 \times 10^{-9}/3.14 \times 1.1 \times 10^{-3}$$
$$= 0.733 \text{ mrad} = 0.420°$$

The divergence angle of the beam is extremely small, so it can reach long distances.

Using Equation 4–9, we can calculate the Rayleigh range thusly:

$$z_R = d_0/\theta = 1.1 \times 10^{-3}/0.733 \times 10^{-3} = 1.50 \text{ m}$$

The task of the optoelectronics designer is to modify the beam diameter, control the divergence or convergence of the beam, and determine the diameter and location of the waist, all of which can be done using lenses or lens systems. Figure 4–5 shows a typical setup with variables to be used in further discussions. It should be noted that the same designation of variables used in the chapter on geometric optics is used here, too, even though there is neither an object nor an image. The object side is considered to be the side where the beam enters the lens; the image side, where it exits.

The Newtonian equation, introduced in Chapter 2, is

$$o'i' = f^2 \qquad\qquad\qquad (4\text{–}10)$$

This equation requires some modification to account for diffraction. Consequently

$$o'i' = f^2 - f_0^2 \qquad\qquad\qquad (4\text{–}11)$$

where o' = distance from entrance focal point to the entrance waist (m)
i' = distance from exit focal point to the exit waist (m)
f = focal length (m)
f_0 = correction for diffraction (m).

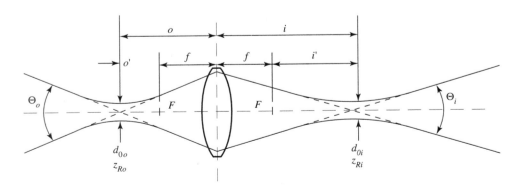

FIGURE 4–5 Transformation of laser beam by a thin lens.

The element f_0 can be expressed in several ways, most commonly as

$$f_0{}^2 = z_{Ro} z_{Ri} \tag{4–12}$$

where z_{Ro} = Rayleigh range on entrance side (m)
z_{Ri} = Rayleigh range on exit side (m).

Applying previously given equations, we can derive the following relationships for f_0:

$$f_0 = d_{0i}/\Theta_0 = d_{0o}/\Theta_i$$
$$= \pi d_{0o} d_{0i}/4\lambda = 4\lambda/\pi\theta_o \theta_i \tag{4–13}$$

where subscript o designates the parameters on the entrance side, and subscript i, parameters on the exit side of the lens.

The relationship between beam waist diameter and its location can also be expressed as

$$(o - f)/(i - f) = o'/i' = d_{0o}{}^2/d_{0i}{}^2 \tag{4–14}$$

Using Equation 4–14, we can develop a simple relationship between the beam parameters on the entrance and exit sides; accordingly

$$d_{0i}{}^2 = f^2 d_{0o}{}^2/[(o - f)^2 + z_{Ro}{}^2] \tag{4–15}$$

Defining a **parameter** α as

$$\alpha = f/\sqrt{(o - f)^2 + z_{Ro}{}^2} \tag{4–16}$$

we can express the relationships between entrance and exit side parameters in a simple form:

$$d_{0i} = \alpha d_{0o} \tag{4–17}$$

$$\Theta_i = 4\lambda/\pi d_{0i} = 4\lambda/\pi\alpha d_{0o} = \Theta_o/\alpha \tag{4–18}$$

$$z_{Ri} = \alpha^2 z_{Ro} \tag{4–19}$$

$$i = f + \alpha^2(\sigma - f) = f + \alpha^2 o' \tag{4–20}$$

Focusing a Laser Beam Using Lenses

Chapter 3 has given us the tools to solve the most common design problems faced by a laser optics design engineer: (1) focusing a laser beam to a minimum waist diameter; (2) collimating the beam or achieving a beam with a long Rayleigh range.

The first problem can be solved in several ways, depending on the location of the lens with respect to the beam waist. When the lens is outside the beam waist, the problem is to refocus an entrance side waist to a new waist on the exit side of the lens. In this case, according to Equations 4–17, 4–18, and 4–19, the parameter α acts as a waist magnifier since $d_{0i} = \alpha d_{0o}$. At the same time, the divergence is decreased by the factor of α, and the Rayleigh range will be multiplied by the factor of α^2.

For a small waist diameter, the parameter α should be minimized. Looking at the expression for α (Equation 4–16), we can see that this can be achieved in two ways: by using a short focal length lens or by placing the lens far away from the entrance waist. The latter increases the lens diameter since the lens must capture the entire beam.

In a case where $o \gg z_{Ro}$, for given entrance and exit waist distance and the desired waist reduction ratio, infinite combinations of focal lengths and geometries are possible. The only limitation in selecting the focal length of the lens is

$$f > f_o = \sqrt{z_{Ro} z_{Ri}} \tag{4–21}$$

Within this constraint, the lens location can be calculated from

$$o = f + \sqrt{(f^2 - f_o^2)}/\alpha \tag{4–22}$$

Application of these relationships is demonstrated in Example 4–3.

EXAMPLE 4–3

A helium-neon laser with $\lambda = 543$ nm and $d_o = 0.84$ mm should be refocused to a waist diameter of 0.2 mm. Design the system.

Solution
From Equation 4–3 we have

$$d_o = 4\lambda/\pi\Theta, \quad \text{or} \quad \Theta = 4\lambda/\pi d_{od} = 4 \times 543 \times 10^{-9}/3.14 \times 0.84 \times 10^{-3}$$

$$= 0.823 \text{ mrad} = 0.047°$$

From Equation 4–8, $z_R = d_o/\Theta = 0.84 \times 10^{-3}/0.823 \times 10^{-3} = 1.02$ m
From Equation 4–17, $d_{0i} = d_{0o}\alpha$ or

$$\alpha = d_{0i}/d_{0o} = 0.20/0.84 = 0.238$$

and from Equation 4–18 $\Theta_i = \Theta_i/\alpha = 0.823/0.238 = 3.45$ mrad $= 0.20°$ and from Equation 4–19 $z_i = z_o\alpha^2 = 1.02 \times 0.238^2 = 0.0578$ m and from Equation 4–12 $f_o = \sqrt{z_o z_i} = \sqrt{1.02 \times 0.0578} = 0.243$ m $= 243$ mm.

Using the criteria from Equation 4–21, the focal length of the lens must be greater than 243 mm. Let's select $f = 300$ mm. Then, from Equation 4–22:

$$o = f_o + \sqrt{(f^2 - f_o^2)}/\alpha = 300 + \sqrt{(300^2 - 243^2}/0.238 = 1039 \text{ mm}$$

Thus, the lens is just outside the Rayleigh range. The minimum lens diameter can be calculated from the entrance divergence and waist distance. Thus

$$D_{\min} = o\Theta = 1039 \times 0.823 \times 10^{-3} = 0.855 \text{ mm}$$

and the location of the reduced exit waist from Equation 4–13 is

$$o'/i' = 1/\alpha^2 \quad \text{or} \quad i' = o'\alpha^2 = (o - f)\alpha^2 = (1039 - 300)0.238^2$$

$$= 41.8 \text{ mm} \quad \text{and}$$

$$i = f + i' = 300 + 41.8 = 341.8 \text{ mm}$$

The entire system is shown in Figure 4–6.

When focusing a collimated beam to a small spot, several simplifications can be made. When the waist distance o and Rayleigh range z_{Ro} is large compared to the focal length f, Equation 4–16 reduces to

$$\alpha = f/\sqrt{o^2 + z_{Ro}^2}$$

There are two possible conditions in this case:

(1) *The lens is well inside the Rayleigh range*, or $o << z_{Ro}$. In this case:

$$\alpha = f/z_{Ro} = f\theta_o/d_{0o} \tag{4–23}$$

and

$$d_{0i} = \alpha d_{0o} = f\theta_o \tag{4–24}$$

$$\theta_i = \theta_o/\alpha = d_{0o}/f \tag{4–25}$$

$$z_{Ri} = \alpha^2 z_{Ro} = f^2/z_{Ro} \tag{4–26}$$

$$i = f + \alpha^2(o - f)f \tag{4–27}$$

(2) In the case where *the lens is well outside the Rayleigh range,* or $o >> z_{0o}$, we can write

$$\alpha = f/o \tag{4–28}$$

$$d_{0i} = \alpha d_{0o} = f d_{0o}/o \tag{4–29}$$

$$\theta_i = \theta_o/\alpha = \theta_o o/f \tag{4–30}$$

$$z_{Ri} = \alpha^2 z_{Ro} = z_{Ro} f^2/o^2 \tag{4–31}$$

$$i = f + f^2/o \approx f \tag{4–32}$$

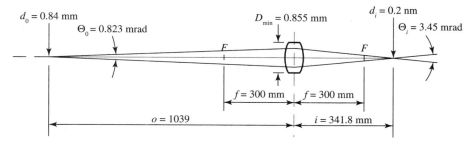

FIGURE 4–6   Solution of Example 4–3

In this case, the lens is far away from the waist and care should be taken so that the diverging beam does not overfill the lens. Therefore, the minimum lens diameter D_{min} should be

$$D_{min} = o\Theta_o \tag{4-33}$$

Example 4–4 illustrates the focusing problem in this case.

EXAMPLE 4–4

Using the same helium-neon laser as in Example 4–2, find the exit waist diameter and location when an $F = 80$ mm lens is used at the following locations:

(A) $o = 0.50$ m. The lens is inside the entrance of the Rayleigh range.
(B) $o = 1.01$ m. The lens is at the limit of the Rayleigh range.
(C) $o = 4.00$ m. The lens is outside the Rayleigh range.

Solution A
Beam parameters are $d_{0o} = 0.84$ mm
$$\Theta_0 = 0.832 \text{ mrad}$$
$$z_{Ro} = 1.01 \text{ m}$$

Applying Equations 4–23 to 4–27, we find

$$\alpha = f/z_{Ro} = 80/1010 = 0.0792$$
$$d_{0i} = f\Theta_o = (80 \times 10^{-3})(0.832 \times 10^{-3}) = 66.6 \times 10^{-6} = 66.6 \ \mu m$$
$$\Theta_i = d_{0o}/f = 0.84/80 = 10.5 \text{ mrad}$$
$$0.0792^2 z_{Ri} = \alpha^2 z_{Ro} = 0.0792^2 \cdot 1.01 = 6.34 \text{ mm}$$
$$i = f + of^2/z_{Ro} = 80 + 500 \times 80^2/1010^2 = 83.1 \text{ mm} \approx 80 \text{ mm}$$

The system is shown in Figure 4–7.

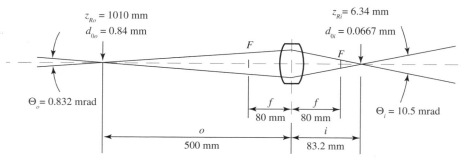

FIGURE 4–7   Solution of Example 4–3(a) $o < Z_{Ro}$

Solution B
In this case Equations 4–16 to 4–20 should be used. Thus

$$\alpha = f/\sqrt{(o - f)^2 + Z_{Ro}{}^2} = 80/\sqrt{(1010 - 80)^2 + 1010^2} = 0.0583$$

$$d_{0i} = d_{0o}\alpha = 0.84 \times 0.0583 = 0.0490 \text{ mm} = 49.0 \ \mu\text{m}$$

$$\Theta_i = \Theta_o/\alpha = 0.832/0.0583 = 14.27 \text{ mrad}$$

$$i = f + \alpha^2(O - f) = 80 + 0.0587^2(1010 - 80) = 83.2 \text{ mm}$$

$$z_{Ri} = \alpha^2 z_{Ro} = 0.0858^2 \times 1010 = 3.48 \text{ mm}$$

The system is shown in Figure 4–8.

Solution C
In this case, Equations 4–28 to 4–32 apply. Thus

$$\alpha = f/o = 80/4000 = 0.020$$
$$d_{0i} = \alpha d_{0o} = 0.020 \times 0.84 = 0.0168 \text{ mm} = 16.8 \ \mu\text{m}$$
$$\Theta_i = \Theta_o/\alpha = 0.832/0.020 = 41.6 \text{ mrad}$$
$$z_{Ri} = z_{Ro} f^2/o^2 = 1010 \times 80^2/4000^2 = 0.404 \text{ mm}$$
$$i = f + f^2/o = 80 + 80^2/4000 = 81.6 \text{ mm} \approx 80 \text{ mm}$$

Since the lens is far away from the entrance waist, minimum lens diameter should also be calculated:

$$D_{\min} = \Theta_o o = (0.832 \times 10^{-3})(4000) = 3.32 \text{ mm}$$

This case is shown in Figure 4–9.

A special case of refocusing the waist is when $\alpha = 1$, *that is, the case of a relay.* In this case, solving Equation 4–15 for o and f, we find

$$o = i = f \pm \sqrt{f^2 - z_{Ro}{}^2} \quad \text{for } f \gg z_{Ro} \qquad \textbf{(4–34)}$$

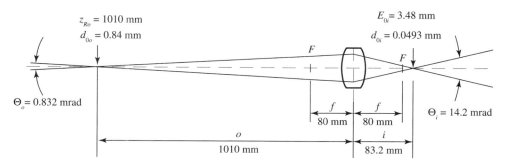

FIGURE 4–8   Solution of Example 4–3(b) $o = Z_{Ro}$

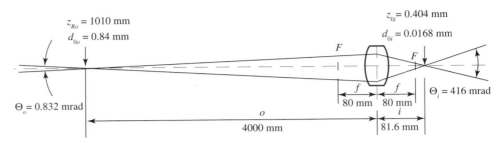

FIGURE 4-9   Solution of Example 4-3(c) $o >> z_{Ro}$

and

$$f = (o^2 + z_{Ro}^2)/2\sigma \qquad\qquad (4\text{-}35)$$

$$o + i = 2f \pm 2\sqrt{f^2 - z_{Ro}^2} \qquad\qquad (4\text{-}36)$$

It is obvious that when $f >> z_{Ro}$, then $o + i = 4f$, just as in the case of normal optics. In the case where $f = z_{Ro}$, however, $o + i = 2f$. This case is called the **Gaussian relay.**

Focusing a laser beam to a collimated beam is opposite of focusing the beam to a small spot. From Equation 4-3 it is clear that a perfect collimated beam with $\theta = 0$ is not possible since it requires an infinite d_0. Therefore we can only approximate a true collimated beam. This can be achieved in two ways: (1) making the diverging angle θ as small as possible; or (2) maximizing the exit Rayleigh range.

The first condition can also be interpreted as maximizing the exit waist distance (i). From Equation 4-20, we can see that i has maximum value when $o = f + z_{Ro}$, in which case

$$i = f + f^2/2z_{Ro} \qquad\qquad (4\text{-}37)$$

To achieve a long exit waist distance, a long focal length lens and a beam with short Rayleigh range should be used. The lens focal point should be at the end of the Rayleigh range of the input waist. This can be achieved by using a two lens system: first lens with a short focal length f_1 to shorten the Rayleigh range and a second lens with a long focal length f_2 to produce a long exit waist distance. Such a system, called the **Gaussian beam collimator,** is depicted in Figure 4-10.

The collimator is usually placed close to the laser; therefore the input lens is inside the Rayleigh range of the entrance beam and Equations 4-23 to 4-27 apply. Accordingly

$$\alpha_1 = f_1\Theta_o/d_{0o}$$

$$z_{R1} = f_1^2/z_{Ro}$$

$$i_1 \approx f_1$$

For maximum collimation, the output lens has to be placed so that $o_2 = f_2 + z_{R1}$. Using Equation 4–16, we obtain

$$\alpha_2 = f_2/\sqrt{2z_{R1}} \tag{4–38}$$

where subscript 2 designates the parameters of the exit lens. The exit waist diameter is

$$d_{0i} = \alpha_2 d_{01} = \alpha_1 \alpha_2 d_{0o} \tag{4–39}$$

Defining

$$\alpha_{12} = \alpha_1 \times \alpha_2 = f_2/\sqrt{2} \times f_1 \tag{4–40}$$

we can see that the parameters of the output beam are

$$z_{Ri} = \alpha_{12}{}^2 z_{Ro} \tag{4–41}$$

$$i = f_2 + \alpha_{12}{}^2 \cdot z_{R0} \approx \alpha_{12}{}^2 \cdot z_{R0} \tag{4–42}$$

$$\Theta_i = \Theta_o/\alpha_{12} \tag{4–43}$$

When collimation by maximizing the Rayleigh range is desired (case 2), the focal point of the second lens should be at the waist of the first image ($f_2 = o_2$) and the following changes in the above equations should be made:

$$\alpha_{12} = f_2/f_1 \tag{4–44}$$

The exit waist is at the focal point of the exit lens:

$$i = f_2 \tag{4–45}$$

and the divergence is reduced by the factor of $\sqrt{2}$. For collimation, a negative and a positive lens can be used also, as shown in Figure 4–10(b).

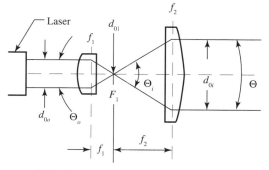

(a) Collimating with two positive lenses

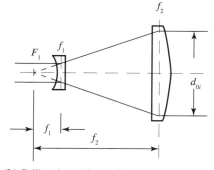

(b) Collimation with a positive and negative lens

FIGURE 4–10   Gaussian beam collimation.

EXAMPLE 4–5

The He-Ne laser in previous examples is often used in surveying instruments and requires a collimated beam for long reach. Design a collimator using a 3.0-mm focal length input lens and a 100-mm focal length output lens. Design the collimator for maximum waist distance and greatest Rayleigh range.

Solution A

For maximum output waist distance $o_2 = f_2 + z_{R1}$

$$\text{Laser: } d_{0o} = 0.84 \text{ mm}; \; \Theta_o = 0.837 \text{ mrad}; \; z_{Ro} = 1010 \text{ mm}$$

Using Equations 4–40 to 4–43 we can write

$$\alpha_{12} = f_2/f_1\sqrt{2} = 100/3.0 \times \sqrt{2} = 23.57, \text{ and the exit waist}$$

parameters are

$$d_{0i} = \alpha_{12}d_{0o} = 2.357 \times 0.84 = 19.8 \text{ mm}$$

$$\Theta_i = \Theta_o/\alpha_{12} = 0.837/23.57 = 35.5 \; \mu\text{rad}$$

$$i = f_2 + \alpha_{12}z_{Ro} = 0.100 + 23.57 \times 1.010 = 23.9 \text{ m}$$

$$z_{Ri} = \alpha_{12}^2 z_{Ro} = 23.57^2 \times 1.01 = 561.1 \text{ m}$$

The Rayleigh range extends over half a kilometer! The beam parameters between the lenses are

$$\alpha_1 = f\Theta_o/d_{0o} = (3.0)(0.837 \times 10^{-3})/0.84 = 2.99 \times 10^{-3}$$

$$z_{R1} = \alpha_1^2 z_{Ro} = (2.99^2 \times 10^{-6})(1010) = 9.03 \times 10^{-3} \text{ mm}$$

$$i_1 = f_1 + of_1^2/z_{Ro}^2 = 3.00 + 20 \cdot 3.00^2/1010^2 \approx 3.00 \text{ mm}$$

$$\Theta_1 = d_{0o}/f_1 = 0.84/3.00 = 0.28 \text{ rad}$$

Minimum output lens diameter is

$$o_2 = f_2 + z_{R1} = 100 + 0.9 = 100.9 \text{ mm}$$

$$D_{2min} = o_2\Theta_2 = 100.9 \times 0.28 = 28.2 \text{ mm}$$

The beam profile is shown in Figure 4–11.

Solution B

Collimating by maximizing the Rayleigh range. In this case, using Equation 4–44, we obtain

$$\alpha_{12} = f_2/f_1 = 100/3.00 = 33.3$$

In this case, the intermediate beam waist should be at the back focal point of lens 2 and the exit Rayleigh range is

$$z_{Ri} = \alpha_{12}^2 z_{Ro} = 33.3^2 \times 1.01 = 1120 \text{ m}$$

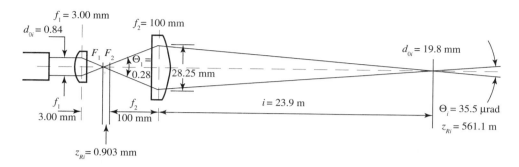

FIGURE 4–11   Example 4–5(a): solution for maximum waist distance.

Depending on the cavity configuration, other laser beams with different energy distribution than the Gaussian or TEM_{00} pattern are possible. Those modes are more complex and generally have a higher dispersion than the TEM_{00} mode.

4–2   PRINCIPLE OF LASER OPERATION

As described in Chapter 3–1, laser radiation originates from electronic transitions in atoms or molecules, involving both absorption and emission of electromagnetic radiation (photons). These transitions can be classified into three types: absorption, spontaneous emission, and stimulated emission. The principles of these phenomena are shown in Figure 4–12.

Figure 4–12(a) depicts **absorption.** In this process, radiation of an energy equal to that of an electronic transition is absorbed, and the atom or molecule is excited to a higher energy level.

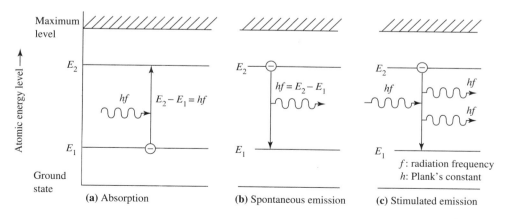

FIGURE 4–12   Interaction between atom energy level and radiation.

Figure 4–12(b) shows **spontaneous emission,** where an electron descends from a higher energy level to a lower level and as a result a photon is emitted. This process takes place in a random manner and is the source of radiation from most common sources.

Figure 4–12(c) explains **stimulated emission.** The reason for radiation is, again, electron descent from higher energy level to lower energy level. This process, however, is not random, but it is triggered by external radiation or a photon that interacts with the atom. The important characteristic of this process is that the released radiation is coherent with triggering radiation: it has the exact same frequency, is in the same phase, and travels in the same direction. In other terms, light amplification takes place. This phenomenon is the basis for laser operation.

In many ways a laser device is very similar to a feedback oscillator. A classic feedback oscillator circuit is shown in Figure 4–13. This is used to explain the operation of a laser.

The heart of the oscillator is the amplifier that produces an output signal, which is an amplified replica of the input signal. The frequency of the signal is determined by the tuned LC circuit. When part of the output signal in proper phase is fed back to the input, a self-oscillating system is created and the output signal will increase to the power limit of the amplifier. A power supply is needed to provide energy for the process.

In a laser the amplification is provided by the stimulated emission process. This takes place in an optical medium that has two or more electron energy levels. A typical two-level medium is shown in Figure 4–14, where level 1 presents the higher energy level and level 0, the lower.

The type of interaction between the medium and applied electromagnetic radiation depends on the electron population in the respective level. When the population in level 0 (N_0) is greater than the population in level 1 (N_1), or $N_0 > N_1$, as in the case of thermal equilibrium, absorption takes

FIGURE 4–13   A feedback oscillator.

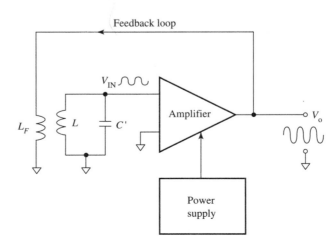

14

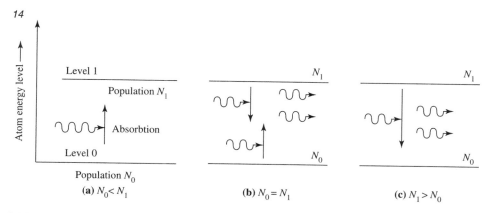

(a) $N_0 < N_1$ **(b)** $N_0 = N_1$ **(c)** $N_1 > N_0$

FIGURE 4–14   Absorption and emission in a two-level system.

place. When $N_1 > N_0$, a condition that is called **population inversion,** stimulated emission takes place and the applied radiation is amplified. In between those conditions is the case where $N_1 = N_0$, called **two-level saturation,** in which the absorption and stimulated emission will compensate for one other and the material looks transparent.

It is obvious that in the two-level system the number of electrons available for stimulated emission cannot exceed the electrons that were absorbed and, therefore, population inversion and amplification in a two-level medium cannot occur. However, in a medium with three or more energy levels, population inversion is possible. A typical condition is depicted in Figure 4–15.

In a three-level system, Figure 4–15(a), electrons are elevated by so-called **pumping action** to higher energy level 2. A characteristic of this level is that the electrons there have a very short lifetime. They decay rapidly to level 1 where they accumulate because the electron lifetime there is much longer. Therefore, an overpopulation in this level occurs. Because of the population inversion, a laser action between level 1 and level 0 can take place.

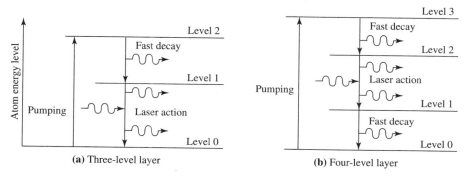

(a) Three-level layer   **(b)** Four-level layer

FIGURE 4–15   Three- and four-level laser systems.

Lasing action can take place also in a four-level system, as shown in Figure 4–15(b). Here, overpopulation occurs between levels 2 and 1–that is, between two levels of very short electron lifetime. For continuous laser operation a four-level system is needed. The three-level system can operate only in a pulsed mode.

Several optical media can support laser action. Ruby crystal, semiconductor junction, and several gases, etc., are typical examples. They are discussed in more detail in Section 4–3.

As mentioned before, an oscillator needs a power supply to provide energy for the oscillation. In lasers the energy is supplied by lifting the electrons to a higher energy level. This is called **pumping**. Pumping can be achieved in many ways: high intensity radiation, high temperature, and electric current, etc. This is also discussed in more detail in Chapter 5.

Finally, an oscillator needs feedback and a frequency determining device. Both are accomplished with a mirror arrangement that encloses the lasing medium, as shown in Figure 4–16.

The lasing medium is between two parallel mirrors, 1 and 2. They direct the beam back through the medium. At each passage, the beam is amplified and the intensity increased. Mirror 2 is semitransparent and allows some of the beam to escape.

Besides feeding energy back to the lasing medium, the two mirrors form a **resonance cavity** or **waveguide** where only one wavelength can oscillate. This gives the laser the characteristic of monochromaticity. The line width of such a waveguide is about six orders of magnitude narrower than the line width achieved by spontaneous emission.

4–3   Types of Lasers

A great variety of lasers with different lasing media, pumping methods, output power, and wavelength are available—too many to cover in depth in this book. Therefore a listing covering only operating principles, output characteristics, and possible applications is attempted in this chapter.

Lasers can be classified by the lasing medium and pumping methods. The first determines the laser wavelength, and the latter, output power and

FIGURE 4–16   Feedback and tuning method in a laser.

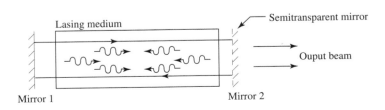

Lasing medium

Semitransparent mirror

Ouput beam

Mirror 1

Mirror 2

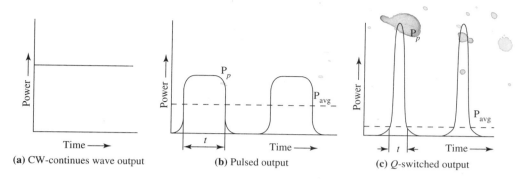

(a) CW-continues wave output   (b) Pulsed output   (c) Q-switched output

FIGURE 4–17   Operating modes of lasers.

operating mode. Lasers can be operated in three modes as shown in Figure 4–17: continuous output, pulsed, and Q-switched mode.

CW or **continuous wave** operation doesn't need further explanation. **Pulsed mode** output is achieved by pulsed pumping schemes. Depending on the pumping methods, repetition rates from a few hertz to kilohertz may be achieved with relatively high bursts of output power.

Q-switched output produces pulses with very short duration, in picosecond or nanosecond range, with peak power in the kilowatt or megawatt range. Q-switching is accomplished by introducing into the resonance cavity a device that lowers its Q^* and thus damps oscillation and prevents stimulated emission. As a result, an excess of electron overpopulation in lasing band is produced. When the oscillating inhibiting mechanism suddenly is removed, a very strong, short duration burst of radiation is generated.

There are several methods for controlling the Q of an optical cavity. The simplest is to make one of the feedback mirrors rotating. This way the cavity can oscillate only during short interval when the mirror faces are parallel. Higher repetition rates may be achieved by inserting electrically or acoustically controlled devices such as **Kerr** or **Faraday cells** or **acousto-optic modulators** into the cavity. These devices can change the polarization of the beam and in combination with a polarizing filter can pass or block the beam.

In this chapter only a short description of the most common lasers can be attempted. They are classified by the lasing medium. Their characteristics are summarized in Table 4–1. Figure 4–18 shows ranges of wavelength and output power.

Two **solid state** lasers, the ruby and the YAG laser, are very common.

The **ruby laser** uses ruby crystal ($Al_2O_3:Cr^{3+}$) as the lasing medium. It was the first material to lase and is still widely used. The ruby rod is pumped with an intense pulse of light produced by a flashbulb, as shown in Figure 4–19.

---

*Q or Q-factor is a term that describes the quality of a resonant cavity or circuit. In a resonant circuit, Q is expressed as a ratio of the reactance of the tuning element to the loss resistance.

TABLE 4-1
Summary of Properties of Common Lasers

Property	Solid State (Ruby, YAG)	Neutral Atom (He-Ne)	Ion Gas (Ar, Kr, Ar-Kr)	Molecular Gas (CO_2)	Dye Laser	Excimer Laser	Laser Diode
Lasing media	Al_2O_3:Cr^{+3} $Y_3Al_5O_{12}$	He-Ne	Ar, Kr, Ar-Kr	CO_2	Dye in solvent	Ar, Kr, Xe	Semiconductor
Pumping method	Flash lamp	HV discharge	HV discharge	HV discharge	Flash lamp laser	HV discharge	Current
Operating modes	Ruby: P YAG: P, CW, Q	CW	CW	CW, P	P	P	CW, P, Mod
Wavelengths (nm)	Ruby: 694 YAG: 946–1320	633, 1150 3390	350–800	$9.6-11 \times 10^3$	300–1000	190–350	670–1550
Beam dimension (mm) (DIA)	Ruby: 1.5–25 YAG: 8–20	0.5–2	0.6–2	1–50	0.2–20	6 × 32 20 × 32 (oval)	0.003–0.007
Beam divergence (rad)	Ruby: 0.2–10 YAG: 0.3–20	0.5–1.7	0.4–2	0.5–10	0.5–5	2–6	0.2–0.5 (oval)
Output power (W)	YAG: 10^{-3}–500	0.1–50 mW	0.5–20	0.1–15,000	0.1–50	1–100	0.01–5
Cooling	Water	Air	Water, air	Water, air	Water, air	Air, water	Air
Price range ($)	3000–150,000	140–15,000	300–50,000	3000–300,000	4000–100,000	30,000–200,000	20–7000
Notes		Least expensive			Most powerful	Tunable	Least expensive

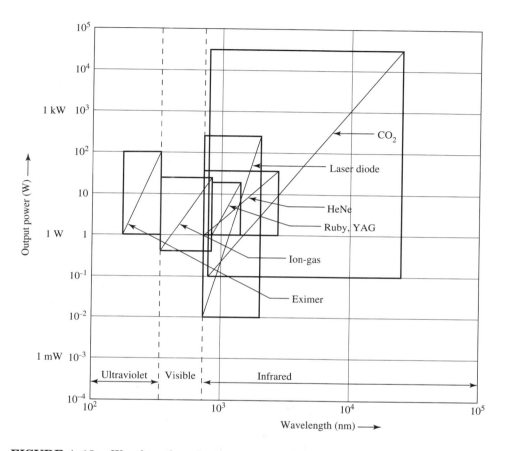

FIGURE 4-18   Wavelength and output, power ranges of common lasers.

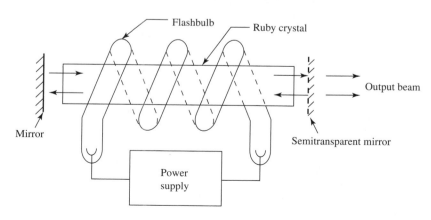

FIGURE 4-19   Ruby laser.

The cavity is formed by two mirrors at the end of the rod or by polished and coated rod ends. The ruby laser is a three-level laser. Therefore it can operate only in pulse mode with pulse duration 10 ns to 10 ms and repetition rate of 0.01 to 4 Hz. It radiates at 694 nm, at the very end of visible red.

The **YAG laser** is the most popular type of solid-state laser. The lasing medium is yttrium aluminum garnet ($Y_3Al_5O_{12}$) crystal, thus the acronym "YAG." It is a four-level laser radiating normally at 1064 nm, but radiations at 946 nm and 1320 nm are also possible. The laser is optically pumped by a flash lamp or a high intensity Xe or Kr lamp. With output power over 500 W, CW and pulse operation, and possible Q-switching, the YAG laser is the most popular type of solid state laser (Figure 4–20).

Solid state lasers are used for general research and development work, holography, materials working, and communications, to name a few fields.

The **gas lasers** are widely used. There are several classes of gas lasers, the **neutral atom laser** being the most popular. The **He-Ne laser** is the most typical of this class. It is very simple and therefore one of the least expensive lasers. Figure 4–21 shows the construction of a He-Ne laser.

The Ne-He laser consists of a sealed glass tube filled with He-Ne gas mixture under low pressure. The pumping is accomplished by two high voltage electrodes inserted into the tube that carry current through the gas, an action similar to a neon display, causing population inversion. The mirrored ends of the tube or separate mirrors provide feedback and cavity action. Lasing takes place at three wavelengths, 633 nm, 1150 nm, and 3390 nm. The

FIGURE 4–20   A YAG laser. (Courtesy Lee Laser, Inc.)

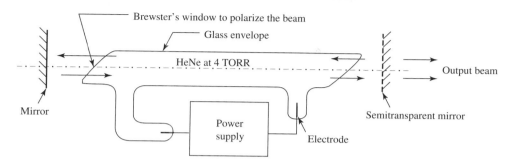

FIGURE 4–21   He-Ne laser.

laser can deliver up to 50 mW at 633 nm and less on longer wavelengths. It operates in CW mode.

Because of its low cost, simplicity, and reliability the He-Ne laser is unquestionably the most widely used laser, with numerous applications in alignment, holography, reprographics, measurement, and many other fields.

Besides He and Ne, other inert gases, such as Ar, Xe, and Kr, can be used in a neutral atom laser. These gases are, however, more common in an **ion gas laser** that works the same way, except with much higher pumping current densities. In these lasers the gas is ionized as the result of high energy collisions. A CW output at higher levels than the He-Ne laser is possible. Emission on several wavelengths can occur, depending on the gas used. Typical output data are:

> argon: 351–528 nm, main line 488 and 514 nm; power 2 mW–20 W
> krypton: 350–800 nm, main line 647.1 nm; power 5 mW–6 W
> argon-krypton: 450–670 nm; power 0.5–6 W.

These lasers are generally built for multiline outputs that are in a visible range. Therefore they are used often for multicolor light shows and displays, besides other applications similar to He-Ne lasers.

The most powerful gas laser group is the **molecular gas lasers, CO_2 lasers** being the most representative in this group. The lasing medium is CO_2 and N_2 mixed with He under low pressure. The gas is circulated between two high voltage electrodes that provide the pumping energy. Depending on the gas flow, in respect to electrode configuration, the devices are classified as transverse and axial flow lasers; see Figures 4–22(a) and (b).

The CO_2 laser can emit wavelengths between 9,000 and 11,000 nm, typically 9,600 and 10,600 nm in infrared range. The pumping scheme in the CO_2 laser is most efficient and the output power, 3 W to 15 kW and higher, is the highest among gas lasers. The laser can operate in CW and pulsed mode. The CO_2 lasers are used for material working (cutting and welding), surgery, laser radar, photochemistry, and many other applications.

FIGURE 4–22 CO_2 laser.

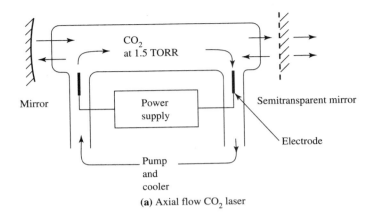

(a) Axial flow CO_2 laser

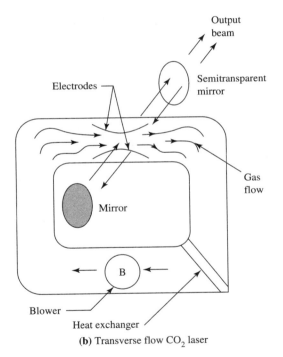

(b) Transverse flow CO_2 laser

All these lasers have one common characteristic—they can operate only at a single wavelength determined by the available energy levels of the lasing media and tuning of the cavity. **Liquid** or **dye lasers** are a group of lasers that are capable of delivering radiation over a wide wavelength range. The lasing medium in these lasers consists of certain organic dye compounds dissolved in ethyl alcohol, methyl alcohol, or water. Depending on the dye selected, a tunable output range from 300 nm to 1000 nm may be achieved.

FIGURE 4–23 A high power CO_2 laser. (Courtesy Rofin-Sinar, Inc.)

When these dyes are pumped by a shorter wavelength using Nd:YAG, or ion laser, or a flash lamp, they exhibit fluorescence emission at a range of longer wavelengths. Using a tunable cavity as shown in Figure 4–24, a continuously varying wavelength from the laser can be obtained.

The tuning is achieved by a rotatable diffraction grating that replaces one of the cavity mirrors. The grating reflects back only one wavelength, depending on the angle θ of the grating. Therefore all other wavelengths are suppressed and stimulated emission is achieved on the selected wavelength only.

The dye lasers can work in CW or pulsed mode. They deliver output power up to 50 W. Their main application is in research and development, medicine, spectroscopy, and fluorescence studies.

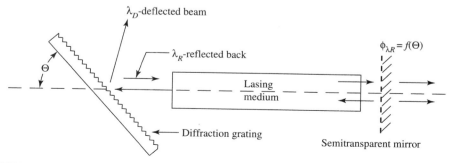

FIGURE 4–24 Principle of a tunable laser.

The **eximer laser** is noteworthy because it can operate in UV range from 190 to 350 nm. The lasing medium in the eximer laser is a mixture of rare gas (Ar, Kr, Xe) with halogen atoms, fluoride or chloride, pumped by an electron beam or electric discharge. The laser operates in pulsed mode with very short pulses (1 to 80 ns) and repetition rates up to 10 kHz. The average output power is up to 100 W, resulting in a very high pulse peak power. The laser is used for general research and development work, photochemistry applications, spectroscopy, medicine, and other fields.

The listing of lasers given above is by no means complete, only the most common and widely used lasers being named. There is a host of other more specialized lasers. Information about these is available in more specific laser literature. The purpose of this listing is only to guide the reader to the proper laser for a given application. More detailed operation principles and application rules should be studied in special literature for a given laser type.

The type of laser most likely to be used by a reader with an electronics background is the laser diode, which is the next to be discussed.

4–4   LASER DIODE

In principle, the laser diode is an LED, with an added optical cavity that provides feedback and generates stimulated emission. At the same time, it also greatly narrows diode radiation bandwidth.

The starting device for the laser diode is the edge-emitting LED, a less common LED than the surface-emitting diode described in Chapter 3. In the edge-emitting diode (see Figure 4–25) the transparent active region is imbedded between two layers. The radiation from this region escapes from the edge of the layers in an elliptical cone pattern.

This configuration is easily adapted for a feedback waveguide by just polishing the radiating sides and reflecting the radiation back to the active region. This type of cavity, however, has a very low Q because of excessive absorption of radiation in p- and n-layers of the diode. The radiation may be restricted to the active region by adding confinement layers to both sides of

FIGURE 4–25 Radiation pattern from an edge-emitting LED.

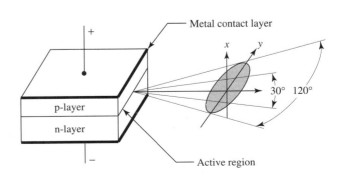

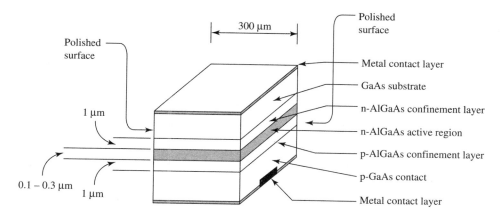

FIGURE 4–26   Laser diode.

the active region, as shown in Figure 4–26. The confinement layers have different refractive indexes, causing the radiation to be reflected back to the active region.

The cavity is completed by polishing the emitting sides. Mirror coating of the sides is not necessary since about 32% of the radiation is reflected back into the cavity anyway, due to the difference in reflective indexes at the air–AlGaAs interface. Radiation therefore takes place from both sides of the device. Often a photodiode is attached to the inactive side, to work as a sensor for the power supply and control the output of the diode. The radiation pattern from a laser diode is similar to the pattern from the edge-emitting LED with one exception—the broad and narrow axes of the elliptical cone are reversed, as shown in Figure 4–27.

With the feedback mechanism in place, the lasing action occurs as soon as the supply of free electrons exceeds the losses in the cavity. The electron

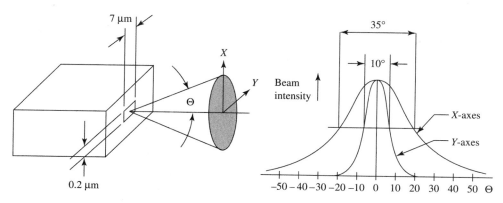

FIGURE 4–27   Radiation pattern from a laser diode.

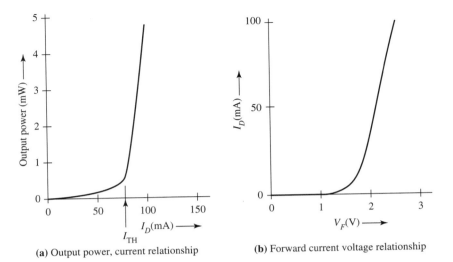

(a) Output power, current relationship   **(b)** Forward current voltage relationship

FIGURE 4–28   Laser diode characteristics.

supply is proportional to the current through the junction. Therefore, the laser diode's minimum threshold current must be exceeded before the lasing takes place. The typical output power and current relationship are shown in Figure 4–28(a). The current voltage characteristic of the laser diode is similar to any semiconductor diode, as seen in Figure 4–28(b). At levels below threshold current, the device acts as a normal LED.

One drawback of the laser diode is its high temperature coefficient. The threshold current increases about 1.5% per °C, as depicted in Figure 4–29. The device has, therefore, the ability to turn itself off after warming up. To avoid this, the laser diode must be operated in a cooled mount or be driven by a constant current power supply, controlled by a built-in photodetector using proper feedback circuit. A typical circuit is shown in Figure 4–30.

FIGURE 4–29   Temperature characteristics of the laser diode.

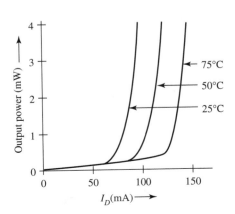

Laser diode power regulation circuit

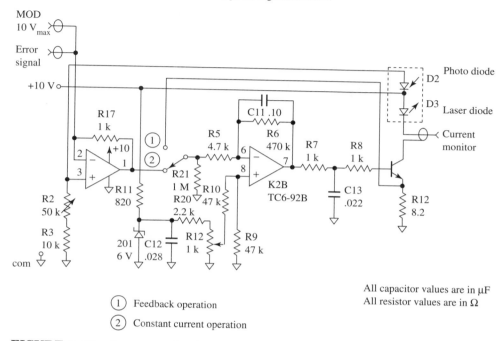

① Feedback operation

② Constant current operation

FIGURE 4–30   Power supply circuit using internal photodetector.

Laser diodes come in many configurations. One that is designed for fiber-optics applications has fiber pigtail for coupling into fiber (Figure 4–31).

The main difference between an LED and a laser diode is the output spectrum. As described in Chapter 3, the radiation in a semiconductor *p-n* junction is the result of electron transfer from a higher energy band to a lower band. The wavelength of radiation is determined by the energy gap between the bands, as given by Equation 3–7.

$$\lambda = 1240/\Delta E \text{ nm}$$

Since the energy gap is fairly constant, the LED should be a monochromatic source. Actually, the LED shows a considerable bandwidth, as shown in Figure 4–32.

The broadening of the bandwidth is caused by the **Doppler effect**—the apparent change in frequency from a moving source. The random velocities of the atoms in the active region give rise to the Doppler effect, producing a range of frequencies centered around the frequency determined by the shift in energy gap. From this semirandom spectrum, the optical cavity selects frequencies at which the cavity can resonate.

In the optical cavity, the light wave is reflected back by the end mirrors. The reflected wave interferes with the approaching wave, either constructively by increasing the combined amplitude when they are in phase or

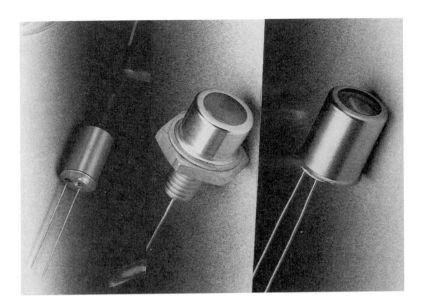

FIGURE 4–31 Typical laser diodes. (Courtesy Laser Diode Products, Inc.)

FIGURE 4–32 Broad
bandwidth from an LED.

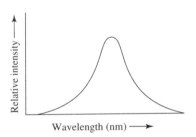

destructively when they are 180° out-of-phase. At certain frequencies a stable standing wave pattern is formed, as shown in Figure 4–33.

At these frequencies the cavity is in resonance, and the oscillation can be sustained with minimum losses. The other frequencies interact destructively and decay rapidly. At resonance, an integral number of half wavelengths of the resonating wave must fit into the length of the cavity:

$$L = m\lambda/2 \quad \text{or} \quad \lambda = 2L/m \qquad \textbf{(4–46)}$$

FIGURE 4–33 Standing
wave pattern in an optical
cavity.

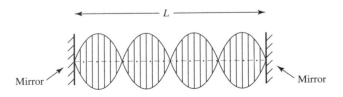

where λ = wavelength of the resonating frequency (m)
L = length of the cavity (m)
m = an integral.

EXAMPLE 4–6

A typical laser diode chip has a cavity length of 300 μm. The AlGaAs has a gap frequency of 800 nm. How many wavelengths will fit into the cavity?

Solution
$m = 2L/\lambda = (2)(300 \times 10^{-6})/800 \times 10^{-9} = 750$

The cavity will resonate with a standing wave pattern of 750 wavelengths. However, it will also oscillate at 748, 749, 751, 752, etc. wavelengths, if the gain of the lasing medium supports these frequencies. As seen from Figure 4–32, the semiconductor medium has a wide bandwidth and, therefore, we can expect a wide spectrum of frequencies emitted from a semiconductor laser such as the one in Figure 4–34.

This type of laser diode is called a **multi-mode laser.** In fact, all laser diodes operate in this mode at low lasing current. Only at higher lasing current does the bandwidth narrow to the bandwidth that limits the other modes of oscillations. The spacing between the adjacent cavity modes can be expressed as follows: the velocity of the light wave in the cavity depends on the reflective index of the medium, so

$$v = c/n \tag{4–47}$$

where v = velocity of the wave in cavity (m/s)
c = speed of light (m/s)
n = refractive index of cavity material.

Using this relationship, the wavelength in the cavity may be converted into cavity resonance frequencies. Thus

$$f = mc/2nL \tag{4–48}$$

The spacing between the adjacent frequencies is

$$\Delta f = c/2nL \tag{4–49}$$

FIGURE 4–34   Radiation pattern of a multimode laser.

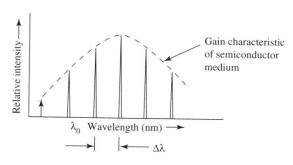

and from this the wavelength spacing

$$\Delta\lambda = \lambda_0{}^2 \Delta f/c \qquad\qquad (4\text{--}50)$$

where $\lambda_0 = $ the free space value of the wavelength (m)

EXAMPLE 4-7

Using Example 4–6 and assuming that the line width of the cavity is 2 nm and the refractive index $n = 3.6$, find the output spectrum from the laser.

Solution
The frequency spacing between modes is

$$\Delta f = c/2nL = 3 \times 10^8/(2)(3.6))300 \times 10^{-6}) = 139 \times 10^9 \text{ Hz}$$

and the corresponding wavelength spread is

$$\Delta\lambda = \lambda_0{}^2\Delta f/c = (800 \times 10^{-9})^2(139 \times 19^9)/3 \times 10^8$$
$$= 0.275 \times 10^{-9} \text{ (m)} = 0.275 \text{ nm}$$

In a 2 nm band, the laser will emit $2/0.275 = 5.5$, or about six bands of frequencies, as indicated in Figure 4–35.

The laser diode described above is a multi-mode laser. There are also methods to produce single-mode laser diodes. These, more expensive devices, are essential in long-range fiber optics communications.

Laser diodes are relatively low output power devices, typically below 100 mW. Therefore, their application is mainly in the fields of communications, optical reading, printing, recording, range finding, etc. Because of their relatively low cost, the impact of the laser diode in these areas has been remarkable.

Among the many applications of laser diodes, their use in fiber-optics communications is most prominent. The fibers can carry a wider bandwidth than any other communication media. For fiber-optics communication, a

FIGURE 4-35 Radiation pattern of an 800 nm laser diode with 2 nm bandwidth.

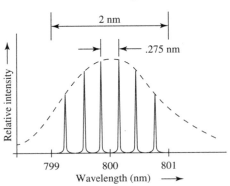

wide band modulation of light beam is needed, and the laser diode works well because of its very fast response and simplicity in modulating the laser beam. The rise time of the laser diode is in the range of 0.1 to 1 ns for a modulation in bandwidth of 2 GHz. The principal mode of modulation, shown in Figure 4–36, is relatively simple.

For fastest response, biasing the diode close to the threshold current is recommended.

Table 4–2 gives a list of available laser diodes and their output frequencies. Table 4–3 compares LEDs, multi-mode lasers, and single-mode laser diodes.

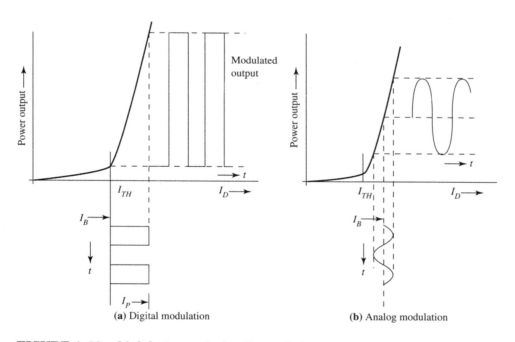

(a) Digital modulation   **(b)** Analog modulation

FIGURE 4–36   Modulation methods of laser diodes.

TABLE 4–2
Available Laser Diodes

Lasing Media	Wavelength (nm)	Line Width (nm)	Power Range (mW)
InGaAlP	670	0.001–4.0	2.0–4.0
GaAlAs	750–850	0.001–15.0	5.0–5000
InGaAsP	1300–1550	0.10–80.0	0.01–100

TABLE 4–3
Comparison of LED and Laser Diode

Property	LED	Multimode Laser Diode	Single-mode Laser Diode
Bandwidth (nm)	20–100	1–5 (80)	<0.2
Rise time (ns)	2–250	0.1–1	0.1–1
Mod. bandwidth (GHz)	<.3	<2	2
Temperature sensitivity	Low	High	High
Lifetime (h)	100,000	10,000–100,000	10,000–100,000
Cost	Low	High	Highest

4–5 APPLICATIONS OF LASERS

The lasers can be used in many applications to perform numerous jobs. Whatever way they are used, all lasers have one common property: **They are very dangerous devices!** Because of the high power concentration, even at milliwatt level, a laser can cause eye injury when observed directly. Therefore, safety precautions described with each laser should be strictly followed. When in doubt, ANSI Standard Z-136.1 or IEC TC-76 should be followed.

The general rules for working with lasers are:

1. Always wear special goggles designed for the given laser.
2. Never look into the primary laser beam or into the specular reflection of a beam.
3. Higher power beams should always be terminated with an absorbing termination and if possible be in a protective enclosure.
4. Skin protection should be worn at higher power levels. A heavy white cloth (such as a standard laboratory smock) will reduce the exposure by a factor of 100.
5. An invisible beam can cause as much damage as a visible beam. Therefore special precaution should be taken when working with wavelengths outside the visible range.

The invention of lasers gave industry and science a wonderful new tool, allowing known operations to be performed more effectively and accomplishing tasks that were impossible with any other technology before the laser. Laser technology is only in the very beginning of its development. New devices and methods appear monthly and this trend is expected to continue for many years to come.

Lasers have had significant impact for military applications and thermonuclear fusion, fields too specialized to cover in this text. Apart from those, laser applications may be divided into four main areas: (1) industrial and material working applications; (2) information processing and optical communications; (3) medical and biological applications; and (4) scientific and metrology applications.

(1) *Industrial and material working applications* is unquestionably the field in which most of the existing lasers are used. Cutting, welding, drilling, surface treatment and alloying of metals are all tasks for which the laser has improved conventional processes in many ways. For metal working, laser beams produce much less heat so material distortion is greatly reduced. Precision is higher than that of conventional cutting methods and more intricate shapes can be cut with lasers than with any other technology. Lasers can achieve higher production rates than conventional methods. Lasers can reach physical areas that are difficult to access with other methods. Materials such as diamond, sapphire, and ceramics that are very difficult to handle with other methods can be cut with lasers. Lasers are excellent for applications on a miniature scale, as required in microelectronics. The laser process is adaptable for easy automation and provides freedom from tool wear. Lasers can improve results of surface treatment of metals. Laser tools are versatile and can be used to manipulate plastics, cloth, leather, wood, ceramics, etc. Resistor trimming, a common laser application, is a typical example. There is literally no end to laser adaptability.

Lasers do, however, have some disadvantages too: the capital cost and running cost in some applications are higher than conventional methods; for reliable operation the laser requires different and sometimes higher-level skills; lasers present new and different safety problems in the workplace.

The higher powered CO_2 and YAG lasers in 100-W to kW range are the mainstay of industrial technology. A typical setup is shown in Figure 4–37.

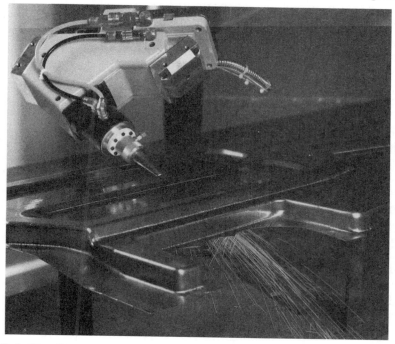

FIGURE 4–37   Typical CO_2 laser metal cutting set-up. (Courtesy NTC Corporation)

(2) In *information processing and optical communications,* lasers have opened new fields without any counterpart in prior technology.

Optical memories first appeared several years ago as compact audio disks (CDs). Information is recorded on the disk by inscribing into the recording material pits one-quarter wavelengths in depth using a laser beam. The pits are arranged in concentric or spiral tracks so that digital information can be stored in the tracks. The tracks are then read by a laser beam concentrated to a very small spot size. He-Ne or laser diodes can be used. The system has two advantages: since the information recorded is in digital form, the noise inherent in any other recording medium is eliminated. Also, an extremely high concentration of information is possible. Disks for video programs and information disks storing entire encyclopedias are available. The same technique for computer application is under development. The compact disk is not subject to any wear when used and therefore has a practically unlimited lifetime.

The laser has found many applications through its ability to scan information. Bar code readers, now so widely used at supermarket checkout counters, are the most common demonstration of this technology. Products carry a cleverly designed code of vertical bars with different widths and spacing. When an He-Ne laser beam is scanned across this code in either direction, the reflections from the bars produce a 10-digit code that identifies the product. A computer using this code can determine the price of the product and make necessary adjustments in inventory.

Laser scanning can also generate images. Laser graphics is the general term for this technique. Laser printers, typesetters, and facsimile transmitters (FAX machines) are typical examples. The images for laser graphics can be generated by a computer or be read from an original by a photodiode, and then be transmitted and regenerated anywhere using laser printers. For example, publishers transmit entire issues of newspapers electronically for printing in different places across the country or world.

The laser beam is an excellent carrier for images because of its very wide frequency bandwidth. Direct use of laser beam through the atmosphere is limited by atmospheric conditions—fog, rain, air pollution, and clouds—but using the optical fiber as a beam carrier has revolutionized the communication industry. A few years ago the first (and already outmoded) fiber-optics cable was laid across the Atlantic. The cable more than doubled all existing communication capacity. A fiber link to every home, capable of carrying all TV channels, computer links, and the telephone communications, will be possible in only a few years. Chapter 8 discusses fiber optics communications technology.

(3) In *medicine and biology,* the laser is used as a surgical or diagnostic tool. In laser surgery, a focused laser beam is used as a scalpel. The infrared beam from a CO_2 laser is absorbed by the water in body cells. The water evaporates, "cutting" the cell. Laser surgery has several advantages compared

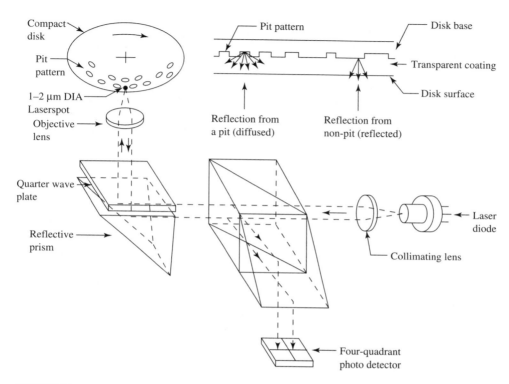

FIGURE 4–38   Laser compact disk reader.

to conventional techniques: incisions can be made with great precision, often directed by a microscope; it is possible to operate in inaccessible areas such as inside an eye or ear and in the gastrointestinal tract; there is limited damage to the adjacent tissues; and bleeding is reduced due to the cauterizing action of the laser beam. In surgery, CO_2 and YAG or Ar lasers are used.

The disadvantages of laser surgery are its complexity, high equipment cost, relatively slow pace, and safety hazards.

As a diagnostic tool, the laser is used for analyzing and counting blood samples, analyzing DNA, and determining the state and condition of cells and biomolecules.

(4) The most exiting and versatile applications of laser technology can be found in the fields of *science and metrology*. From the simple He-Ne laser beam directed to a rotating mirror for establishing an accurate horizontal reference line in new construction sites, to the most precise and accurate measurement and alignment instrumentation, the laser has mostly replaced optical instruments. The perfectly straight laser beam directed to a simple four-quadrant photodetector, allows alignment accuracy of approximately 25 μm over 25 m—satisfactory for the most demanding jobs.

For distance measurement, laser techniques provide high accuracy and simplicity. For shorter distances, typical in machine shop measurements, the interferometer principle is used—the laser beam is reflected back from a mirror. When the reflected beam interferes with the original beam, the interference fringes at the placement of half wavelength of the beam can be observed. By counting the fringes, the movement of the reflecting mirror can be measured to within a half wavelength of the beam. The practical instrumentation produces accuracy on the order of 1 part per million, or 1 μm in a 1 m distance.

Over longer distances, the beam is amplitude modulated, and the phases of the original and reflected beams are compared. This method is also accurate to about 1 ppm or 1 mm over 1 km. For greater distances, short Q-switched pulses are used, and the traveling time of the returned pulse is measured. This method has been used for ranging satellites and measuring the distance between the moon and the earth. The latter measurement is so precise (within 20 cm) that it can be used to measure the shift of the geodectonic plates between continents.

Because the laser beam is highly monochromatic, a very small Doppler shift of the returning beam from a moving target can be accurately measured. This measurement is the basis for developing instruments to measure velocity, where speeds from a fraction of a meter to hundreds of meters per second can be determined. Since the method requires no contact with the measured object, the speed of very small particles can be measured.

Based on the Doppler effect of a moving beam, an interesting laser accelerometer or angular velocity detector has been developed for use as a gyroscope in navigation systems.

The monochromatic nature of the laser beam makes it very effective for measuring scattering, either of small particles or of molecular structures. A laser system can measure the concentration of air pollution, smoke, and gases, since they all have characteristic scattering patterns. Usually ruby, YAG, and infrared lasers are used for this purpose.

These devices represent only a small sample of laser applications in the fields of metrology and science. One more, **holography,** a technique for creating three-dimensional images of an object, deserves special attention. (See Figure 4–39.)

The holographic image, or **hologram,** is recorded by illuminating an object with a laser beam. The beam is directed to the object through a semi-transmitting mirror. The mirror reflects part of the beam to a photographic plate, forming an interference pattern from the beam scattered by the object. On the plate this fine interference pattern is recorded. This pattern is not a photograph and has no visual resemblance to the object, but does contain a complete record of the object.

To view the hologram, the object is removed and the hologram is exposed only to the laser beam from the mirror. As the beam passes the

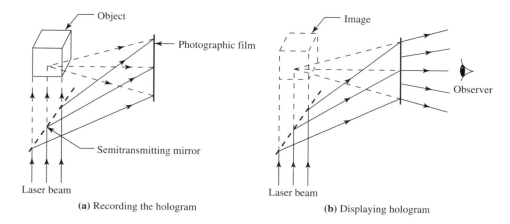

(a) Recording the hologram

(b) Displaying hologram

FIGURE 4–39   Holography.

hologram, the same interference pattern as that of the original object is created, and the observer sees a true three-dimensional replica of the object, with perspectives that change as the observer's position moves.

Holography has many practical applications. An interesting one is holographic interferometry—recording small displacement patterns in an object caused by stresses or vibrations. The technique is quite simple. Two holographic images are recorded, one with and the other without stresses. The small displacements resulting from stress or vibration create interference patterns visible in the images, which can be analyzed.

Finally, researchers are hard at work developing holography for adaptation to the storage capacity of computer memory.

SUMMARY

Chapter 4 starts with a short introduction to laser optics of the Gaussian laser beam. Three-beam parameters, divergence and convergence, beam waist, and Rayleigh range are defined and their relationships are shown. The foundation for practical use of existing lasers is laid: Methods and relationships for controlling the beam with lenses are introduced; and methods for focusing the beams to a small waist diameter and for collimating or achieving beams with a long Rayleigh range are shown.

Laser operating principles are explained. Then, the most common laser types are introduced and their operating principles and technical characteristics are explained, including solid state, gas, liquid or dye, and eximer lasers. A comparative table with primary laser characteristics is given. A more detailed description of the laser diode, including some application principles, is given.

In conclusion, a short description of laser uses in industry, communication, medicine and biology, and in science and metrology is presented.

RECOMMENDED READING

O'Shea, D. C. 1985. *Elements of Modern Optical Design.* New York: A Wiley-Interscience Publication, John Wiley & Sons.

Hecht, J. 1992. *Understanding Lasers.* New York: IEEE Press.

Charscham, S. S. 1972. *Laser in Industry.* New York: Van Nostrand Reinhold Company.

Siegman, A. E. 1986. *Lasers.* Mill Valley, CA: University Science Book.

Buying Guide 1990. *Lasers and Applications Magazine.* Torrance, CA: A High Tech Publication.

PROBLEMS

1. Irradiance at the center of a Gaussian laser beam is 500 mW/mm^2 and the beam width is 2 mm. Find the irradiance at 0.5 mm from the beam center.

2. A Neodymum-YAG laser operating at 1.064 ηm has beam divergence of 0.03°. Find the waist diameter and Rayleigh range.

3. For the laser in Problem 2, find the beam diameter at 10 cm and 75 cm from the waist.

4. A 50 mm focal length lens is located 250 mm from the entrance of a laser beam with the following entrance parameters: $\theta_o = 5.20$ mrad; $Z_{Ro} = 0.157$ m; $d_{0o} = 0.820$ mm. Find the minimum lens diameter and the same beam parameters at the exit side of the lens.

5. A He-Ne laser that operates at $\lambda = 633$ mm and has a waist diameter of $d_0 = 1$ mm is used to read a bar code 300 mm away. Desired $d_{0i} = 0.1$ mm. Design the system.

6. A He-Ne laser that operates at 633 nm and has $d_0 = 0.8$ mm is used in a pointer that projects a spot to a screen 10 m away. Design the system and find the spot diameter.

7. To measure absorption in the atmosphere, a He-Ne laser beam operating at 633 nm and having a waist diameter of 0.5 mm is directed to a receiver 0.5 km away. Design the system using two positive lenses and find the beam diameter at the receiver. Try a solution in which the waist is placed in the middle of the laser—receiver distance.

8. Design a collector lens for the system described in Problem 7 to reduce the beam diameter at the receiver to 0.1 mm.

9. For the given applications, select a laser and give reasons for your selection.
(1) compact disk reader
(2) car chassis alignment device
(3) metal cutting device.
10. List three important safety precautions when working with lasers.

The answers are in Appendix C.

5
Displays

We are living in the age of machines—some sophisticated technology, indeed. The time when devices had only an ON/OFF switch or a pilot light to indicate their status has long passed. Nowadays, using machinery may require complex operations, and frequently a machine will direct us how to operate it. Often, the machine's output is a set of data or information. The machine communicates with us, and we have to communicate with it in return.

Such communication requires using our auditory and visual senses, the latter in the form of reading displays or printed records. Sometimes multiple ways of machine communication are used simultaneously, as in modern cash registers where the result of laser scanning of a bar-code is reported back in voice, visual display, and printed record. By far the most common method is the visual display. In fact, display technology is one of the most common applications of optoelectronics.

In displays, optoelectronics sources are used to convey information. All sources covered in Chapter 3 are prime elements for displays. Interestingly, however, the most popular display does not use a light source. The **LCD,** or **Liquid Crystal Display,** uses an absorber, instead of an emitter, to convey information. Therefore LCD devices do not fit under source classification and are covered separately in this chapter.

5–1 THE LIQUID CRYSTAL DISPLAY

Optically, materials can be classified as transparent, opaque, or all degrees between these two extremes. This classification depends on the extent to which the material reduces the intensity of light, or the amplitude of electromagnetic oscillation, passing through it. Transparent materials allow light to pass through them without reducing the amplitude, while opaque materials do not.

There is another class of materials that produces the same visual result as opaque materials, but works on a different principle. They are called **polarizers.** When a light wave of circular or random polarization passes through a polarizer, its energy is also reduced, not by reducing its amplitude, but by absorbing the waves of certain polarization. This phenomenon is illustrated in Figure 5–1.

On the left in the figure is a **nonpolarized light source;** that is, the amplitude of the oscillation swings in all directions, perpendicular to the propagation of the wave. This is also called a source with **circular polarization** [Figure 5–1(a)]. When this radiation passes through a vertical polarizer, the oscillations in the horizontal direction are absorbed and a vertically polarized wave [Figure 5–1(b)] results. Since one-half of the wave is absorbed, the flux of the transmitted wave is less than the flux of the incident wave. The polarizer is semitransparent. When the same wave now passes through a second polarizer which is horizontally oriented, the waves of vertical polarization are absorbed and the result is that no or very little light energy passes through both polarizers. The combination of two polarizers that are perpendicularly oriented looks opaque and blocks all light [Figure 5–1(c)]. Interestingly, when we change the direction of polarization of one of the polarizers (the second one shown in Figure 5–1) to vertical, it will pass the wave from the first polarizer, and the combination appears semitransparent again [Figure 5–1(d)].

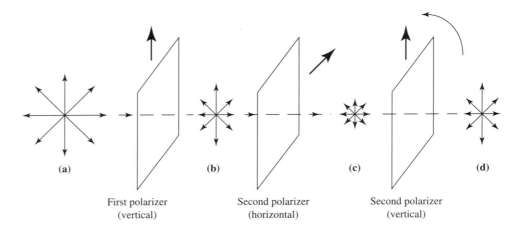

(a)   (b)   (c)   (d)

First polarizer   Second polarizer   Second polarizer
(vertical)   (horizontal)   (vertical)

FIGURE 5–1   Lightwave of random polarization passing through two perpendicularly polarizing filters.

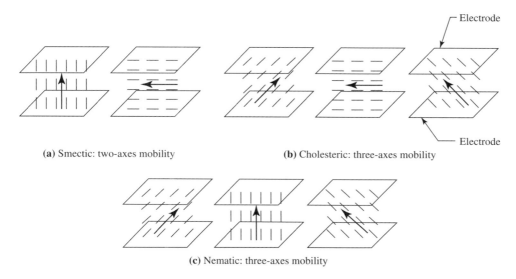

(a) Smectic: two-axes mobility   (b) Cholesteric: three-axes mobility

(c) Nematic: three-axes mobility

FIGURE 5–2   Types of liquid crystal materials.

This technique of using two polarizers and changing the orientation of one to build a light switch is the underlying principle of liquid crystal displays.

Crystals with an aligned molecular structure are used as polarizers. Some organic liquid materials exhibit the same properties. These are called **mesomorphic;** that is, they have a phase between the solid and liquid states where the material is **anisotropic.** In this state, the material exhibits different refractive indexes for the parallel and perpendicular axes, very similar to that of a crystal. What makes these materials usable as displays is that their molecules are not in fixed arrays, as in solid crystals, but can be aligned by an electric field. This characteristic is called **electrical anisotropy.**

There are three types of liquid crystals: **smectic, nematic,** and **cholesteric,** shown in Figure 5–2.

The molecules in smectic material can be aligned in two directions, cholesteric and nematic, in three directions as shown. Most interesting is the **twisted nematic material** that is often used. The construction and operation of a twisted nematic liquid crystal cell is shown in Figure 5–3.

A very thin layer (about 8 μm thick) of liquid crystal material is sealed between two glass plates. On the inside surface of the plates are placed transparent electrodes in the shape of the desired image. The cell is situated between two perpendicularly oriented polarizers that prevent the light from passing through the package.

By utilizing the liquid crystal action, the polarization of the passing light can be changed and the light transmission controlled. The crystal can take one of two states. With no voltage applied to the electrodes, they are in a **twisted nematic state.** The upper electrode is polished in such a manner

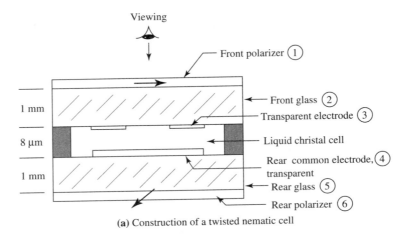

(a) Construction of a twisted nematic cell

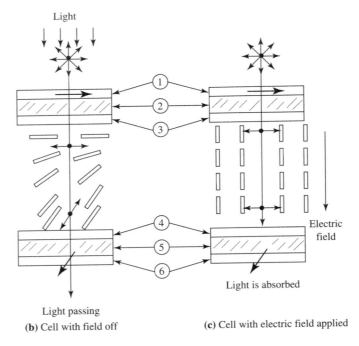

Light passing
(b) Cell with field off   (c) Cell with electric field applied

FIGURE 5–3   Twisted nematic liquid crystal cell.

that the molecules in the upper layer are parallel with the polarization of the upper polarizer. The lower electrode is prepared to align the lower layer of crystal molecules with the lower polarizer. In between, the molecules make a twist from one orientation to another, as shown in Figure 5–3(b). In this condition, the crystal turns the polarization of passing light by 90°, and the light can pass through the lower polarizer. The cell is semitransparent.

With voltage applied to the electrodes, the electric field between the electrodes changes the orientation of crystal molecules, as shown in Figure 5–3(b). With the molecules aligned along the electric field, no change in polarization of the passing light takes place, and therefore the light that tries to pass through the electrodes is blocked. A dark image in the shape of the electrodes is created. If the front and rear polarizers are in the same orientation, a reverse image (that is, a white image on a black background) is produced.

Figure 5–4 shows three ways this technique can be applied to practical displays.

Transmissive display (a) requires a light source behind the display. It is usable under very low ambient light conditions. The reflective display (b) depends on ambient light that is reflected back by a reflector at the rear of the display for its operation. Therefore, it cannot be used in very low ambient light conditions. Watch displays are an example of this type. The third type (c) is a combination of both and is, therefore, usable over a wide range of ambient illumination.

Compared to the displays with active sources, the LCD has several advantages:

1. Very low power consumption with a typical current about 20 nA/mm^2. This makes the LCD applicable for battery-driven devices.
2. Low voltage requirement, typically 1.5 to 5 V.
3. CMOS compatibility.
4. Readability under bright sunlight.
5. Flexibility—the display is easily adapted to segmented, dot-matrix, graphs, graphic displays , and other applications. (Color capability for flat TV displays is also available.)

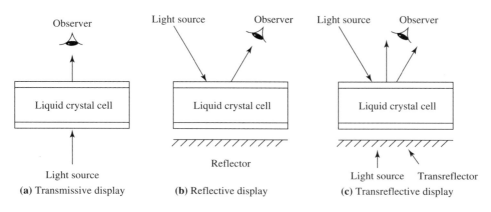

(a) Transmissive display   **(b)** Reflective display   **(c)** Transreflective display

FIGURE 5–4   Modes of liquid crystal dislpay.

LCDs have some disadvantages:

1. The most popular, transreflector type is not usable at low ambient illumination levels.
2. The response time of an LCD cell is too slow for many applications.
3. The viewing angle of the display is limited.
4. These devices are temperature sensitive.

Liquid crystal cells respond to DC as well as AC voltage. Because of electrolysis, the DC voltage causes deterioration. Therefore, either sinusoidal or square wave AC with a minimum DC component is normally used. The cell response to applied voltage is shown in Figure 5–5.

As seen from the graph, the voltage required to activate the cell is dependent on the temperature. At lower temperature, a higher voltage is required.

The contrast of the display greatly depends on the viewing angle. Therefore, the required activating voltage is also a function of the viewing angle, as shown in Figure 5–6.

The viewing angle is defined in Figure 5–6(a). Figure 5–6(c) shows a typical contrast map of an LCD display. The center area with the highest contrast is the best viewing direction. It is slightly off the perpendicular, toward the alignment of the polarizers. The required actuating voltage also depends on viewing angle. The least voltage is needed to form an image when viewed at 45°; the most voltage is needed to form a high contrast image when viewed perpendicular to the display surface ($\alpha = 0°$) [Figure 5–6(b)].

The response time of an LCD cell is slow and temperature dependent. It varies greatly, depending on the cell construction and the liquid crystal material used. Figure 5–7 illustrates a typical example of two commercially available displays. For design purpose, manufacturers' data for a given display should be used.

FIGURE 5–5
Transmission/voltage characteristics of a twisted nematic cell.

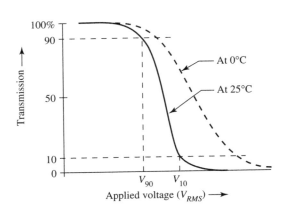

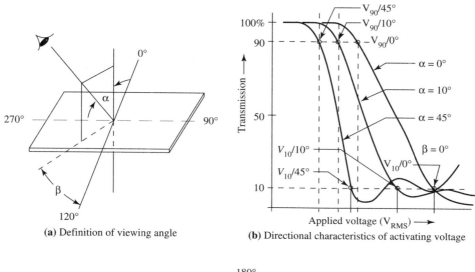

(a) Definition of viewing angle

(b) Directional characteristics of activating voltage

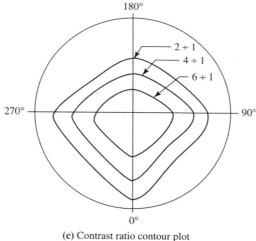

(c) Contrast ratio contour plot

FIGURE 5–6   Directional characteristics of a liquid crystal cell.

5–2   HOW TO DRIVE AN LCD DISPLAY

The equivalent circuit of a liquid crystal cell is a capacitor with a parallel leakage resistance. The values of those components depend on the area and construction of the cell. The resistance values are usually much higher than the reactance of the capacitor. The display is driven with a squarewave or AC. The RMS amplitude should exceed the V_{10} voltage (V_{90} voltage in the case of a reversed display). These values are shown in Figure 5–6(b). The waveform

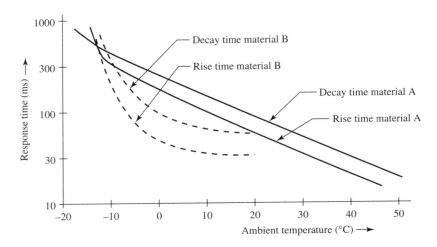

FIGURE 5-7   Typical response time of an LCD display.

should be symmetrical with less than 50-mV DC component to avoid degradation of the liquid crystal liquid. The frequency should be higher than 30 Hz to avoid flicker (up to 1 kHz is common). A higher frequency requires more current because of the capacitive load.

Direct or multiplexed operation may be used. Direct drive is simpler but requires a contact for every element of the display and is therefore used only with less complex displays. A typical squarewave driving circuit for a 7-segment display is shown in Figure 5–8.

The display has an electrode for each segment and a common backplane electrode. A squarewave of proper amplitude is applied to the backplane. The segments are connected to the squarewave source through an exclusive-OR gate. The segments are activated by applying a HIGH to the other input of the gate. The exclusive-OR gate inverts the output when the HIGH is applied. The output is in phase when LO is applied. Therefore, a waveform with an amplitude of $2V_{\text{P-P}}$ is on activated segments and no voltage exists on deactivated segments. Since the liquid crystal cell responds to the RMS value of the waveform, it is important to know that the value of a squarewave is

$$V_{\text{RMS}} = V_{\text{P-P}}/2 \qquad\qquad (5\text{–}1)$$

where V_{RMS} = RMS value of the squarewave (V)
    $V_{\text{P-P}}$ = peak-to-peak value of the squarewave (V).

In addition to the exclusive-OR gates, appropriate decoder and drivers are needed. They are all included in popular LCD driver-decoder packages for one to four digits, such as National DM7211, MM54HC4543, and Intersil ICM7211.

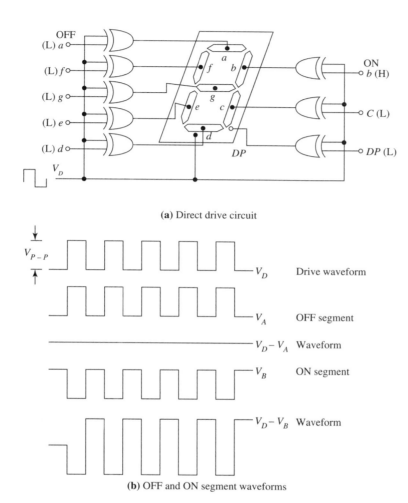

(a) Direct drive circuit

(b) OFF and ON segment waveforms

FIGURE 5–8   7-segment direct drive.

For more complex displays, multiplexed drives are used. The principle of multiplexing is described in this chapter, pp. 179. There, the techniques peculiar to LEDs are covered. The principal difference between LCD and other displays is that the other displays can be turned ON and OFF by just applying or not applying voltage to an element; that is, using a simple pulse waveform. This type of waveform, however, is not suitable for LCDs because it has a DC component. Therefore, the ON pulse in an LCD display should be balanced with appropriate voltage during the OFF period, as shown in Figure 5–9.

Because an LCD responds to the effective, or RMS, voltage of the applied waveform, for proper operation the RMS value of the OFF period should be less than V_{90} and ON voltage should be greater than V_{10}. The fact that

FIGURE 5–9 Driving pulses for regular and LCD displays.

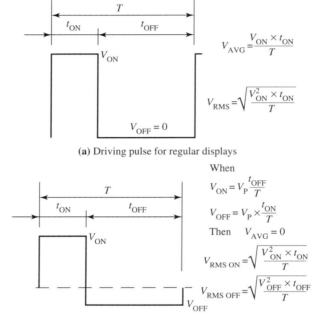

$$V_{AVG} = \frac{V_{ON} \times t_{ON}}{T}$$

$$V_{RMS} = \sqrt{\frac{V_{ON}^2 \times t_{ON}}{T}}$$

(a) Driving pulse for regular displays

When

$$V_{ON} = V_P \frac{t_{OFF}}{T}$$

$$V_{OFF} = V_P \times \frac{t_{ON}}{T}$$

Then $V_{AVG} = 0$

$$V_{RMS\ ON} = \sqrt{\frac{V_{ON}^2 \times t_{ON}}{T}}$$

$$V_{RMS\ OFF} = \sqrt{\frac{V_{OFF}^2 \times t_{OFF}}{T}}$$

(b) Driving pulse for LGD display

these voltages depend on temperature and viewing angle [see Figure 5–6(b)] makes the design of an LCD circuit quite critical and complex.

There is no single approach to a proper design. A compromise between the complexity of the drive circuit and overall performance has to be made. An example of possible drive waveforms is given. It should provide an idea of possible waveform. For all practical purposes, for the average designer of LCDs, they are too complex to generate. There are ready-made LSI chips available to generate these waveforms. Understanding the waveforms, however, is beneficial for all LCD designers.

The example is given in the form of a matrix drive, with N rows and M columns. The rows represent backplanes; the columns represent cells in each backplane. For example, a 4-digit 7-segment display has four rows (backplane for each digit) and eight columns (seven segments plus decimal point). The examples are shown in Figure 5–10.

The example uses four levels of voltage drives. As you can see, they are not quite balanced for zero DC component. To eliminate the residual DC component, two different frame cycles are used alternately. The pulse polarity of one cycle is the opposite of the other cycle. Together they generate two opposite polarity residual components that cancel each other.

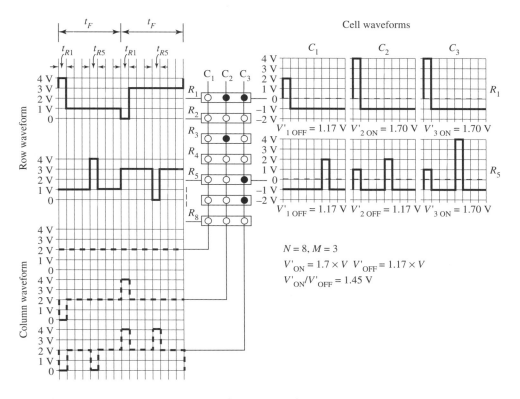

FIGURE 5–10   A four-voltage level 8-row, 3-column display drive.

The object of the design is to achieve the highest V_{ON}/V_{OFF} ratio, giving the display the widest latitude in temperature and viewing angle. There is, however, a theoretical limit that is a function of the number of rows N, given in Table 5–1.

TABLE 5–1
Theoretical Limit of Maximum V_{ON}/V_{OFF} Ratio

N	V_{ON}/V_{OFF}
2	2.41
3	1.93
4	1.73
8	1.45
16	1.29

The voltage steps can be generated within the driver chip or externally using a simple voltage divider, as shown in Figure 5–11.

To compensate for high temperature sensitivity, a negative temperature coefficient resistor can be used as the dropping resistor R_D. The same result can be achieved using a power supply with a positive temperature coefficient.

The waveforms used to drive an LCD display look complicated. Actually, they are easily produced with modern LSI chips. Following are some of the chips that may be used:

- For 7-segment display: Intersil ICM7231, -32, -33; MEM M6232, -33, -34; National COP472, etc.
- Dot matrix display: Hitachi HD44780, 44100; Hughes H0551, Ho551; NEC NEC7227/8, etc.
- General purpose drives: Hughes H0438A; National MM5452/3, MM58438; MEM E078, etc.

5–3   HUMAN FACTORS OF A DISPLAY

A display is the interface between machine and human. It is the means by which we communicate with the machine. Display-machine interaction is usually electronic. Display-human interaction is visual. To be effective, the display must meet certain visual criteria. Three factors are critical for a good visual display: (1) legibility, (2) brightness, and (3) contrast.

Legibility is the property of an alphanumeric symbol that makes it easy to read with speed and accuracy. Two factors contribute to the legibility of a symbol: its style and its size.

FIGURE 5–11   Voltage divider for a multilevel LCD drive.

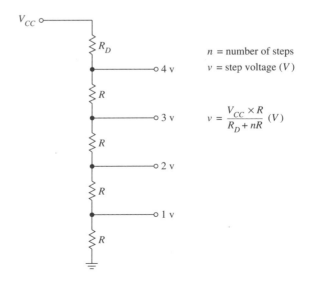

n = number of steps

v = step voltage (V)

$$v = \frac{V_{CC} \times R}{R_D + nR} \ (V)$$

With printed symbols, a great number of font styles with fine resolution and high clarity are available. With electronics, however, fine resolution is costly. Therefore, symbols and characters are produced electronically with very few elements, bars, or dots. Electronic symbols do not create perfect letters and numerals. In situations where speedy recognition of symbols is required, a display with capacity for more clarity (more elements) should be selected.

Displays utilizing small characters should be considered carefully, as small characters may be beyond the resolution limit of our eye, which is about 15 minutes of arc, defined as

$$\tan \alpha = H/D \tag{5–2}$$

where α = viewing angle (°)
   H = character height (m)
   D = viewing distance (m).

The equation can also be written in the form

$$\alpha = 3600H/D$$

where α = viewing angle in minutes.

We re-emphasize that the maximum resolution of the human eye is about 15 minutes; consequently, for different displays, somewhat higher limits are recommended, as follows:

LED display	20 minutes
LCD transmissive display	26 minutes
LCD reflective display	30 minutes

Recommended character proportions are:

Width to height	50 to 100%
Spacing to height	26 to 63%
Segment width to height	13 to 20%
Maximum viewing angle to normal	0 to 19°

EXAMPLE 5–1

Instructions to operate a copier are read from a distance of 30 inches. What is the minimum character height when LEDs are used?

Solution
For LED, the minimum viewing angle is 20 minutes. Therefore:

$$H = \alpha \times D/3600 = 20 \times 36/3600 = 1/6 \text{ in.} = 25.4/6$$
$$= 4.2 \text{ mm}$$

Displays with 3.8, 4.9, and 6.9 mm heights are available. Select 4.9- or 6.9-mm high display.

Other factors such as character sharpness and shape are also significant. Despite difficulties in assigning quantitative values to those qualities, sharpness and shape should be considered, especially in those conditions where it is important for a display message to be comprehended swiftly.

Brightness is our perception of luminance. Brightness determines our perception of the visual world and is the most important element in our vision. Unfortunately, it is a psycho-physical phenomenon extremely difficult to measure objectively or technically. We can, however, easily measure luminance, the stimulus of brightness. A typical luminance measuring instrument is described in Figure 5–12.

The instrument projects the image of the object (in this case, a segment of an LED) to the imaging plane, where it can be observed through an eyepiece. A fiber-optic probe is placed at the area under investigation. The probe measures the illumination of the area, which is then converted to object luminance by multiplying by the square of the magnification.

The luminance/brightness relationship is complex. It depends on luminance level, object color, and other factors that are difficult to determine. Fortunately, the readability of a display does not depend on the brightness of the character but rather on the ratio of its brightness to the background brightness, called the **contrast ratio.** This ratio can be determined by luminance measurements.

Contrast is defined in two ways. For passive displays, such as LCD displays, the contrast is defined as

$$C = (L_O - L_B)/L_O \qquad\qquad (5\text{--}3)$$

where C = contrast
L_B = background luminance (cd/m^2)
L_O = object or source luminance (cd/m^2).

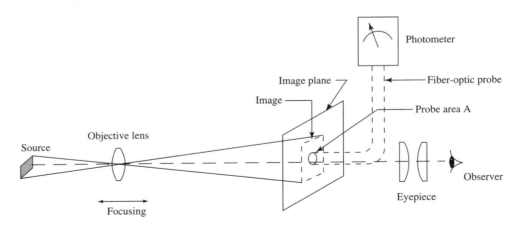

FIGURE 5–12   Apparatus for luminance measurement.

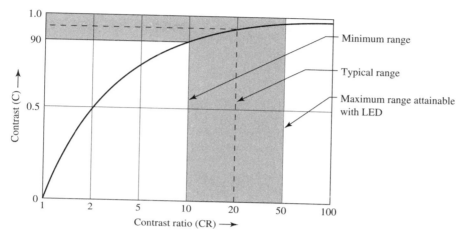

FIGURE 5–13   Relationship and range of contrast and contrast ratio.

For active displays, such as LED and others, the **contrast ratio** is used:

$$CR = L_O/L_B \qquad\qquad (5\text{–}4)$$

where CR = contrast ratio
L_O = object or source luminance (cd/m²)
L_B = background luminance (cd/m²).

The contrast can have a value between 0 and 1, or $0 > C > 1$, zero being the case where object and background luminance are the same and where background has zero luminance.

The contrast ratio can have values between one and infinity, $1 > CR > \infty$. At value one, the object and background have the same luminance and therefore the display is not visible at all; at infinity, background luminance is zero and the display has the best visibility. The relationship between contrast and contrast ratio with the recommended limits is shown in Figure 5–13.

The task of the display designer is to design a display with maximum contrast or contrast ratio. Low contrast values reduce the visual acuity of the eye and therefore reduce the readability of the display. The minimum acceptable contrast and contrast ratios are indicated in Figure 5–13.

To increase the contrast, in the case of a passive LCD display, the background luminance should be increased, either by raising the illumination or by selecting a transmissive display. To increase the contrast ratio of an active display (that is, an LED), the opposite is required: the background luminance should be decreased. Such reduction can be achieved by lowering the ambient illumination through the use of hoods, shuttered covers, and other similar devices, or by recessing the display.

One factor that makes achieving the right contrast more complex than merely adjusting the luminance ratio is the display and background color.

Selecting a proper background color can enhance readability even at low contrast ratios. Filters are the best method for taking advantage of display and background colors for increased readability.

The filter, either band-pass or low-pass, transmits the wavelength emitted by the display and blocks all other wavelengths. Thus, the background luminance is reduced and the contrast ratio increased. A great variety of glass and plastic filters is available for almost all emitters. Polarizing filters can be used to reduce glare and background reflections. Most active displays use filters.

5–4   BAR GRAPHS AND SEGMENTED DISPLAYS

Because of their simplicity and cost effectiveness, bar graphs, and especially segmented displays, are the most widely used. Segmented displays are used in clocks, watches, digital thermometers, and numerous electronic and automotive instruments. The bar graphs are replacing analog instruments as indicators, level meters, and so forth, wherever an analog presentation of a certain quantity is required.

Both displays can be designed and constructed using LCDs or any source described in Chapter 3. Since the design principles for all sources are generally the same, design examples using only LEDs are shown. Using other sources, the same circuits with appropriate drivers can be used.

Bar graphs can be classified in two groups: **bar graphs** and **position indicators.** Figure 5–14 explains the difference.

The bar graph produces a bar composed of several elements. The length of the bar represents the quantity being measured. With the position indicator, only the last element of the bar is displayed. Its position represents the quantity. The principal circuit for both devices is shown in Figure 5–15. The circuit is for a voltmeter that measures voltages from 0 to 10 volts in 1-volt steps.

FIGURE 5–14   Bar graph and position indicator.

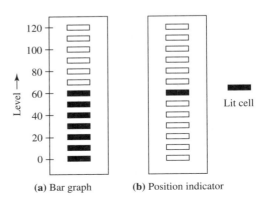

(a) Bar graph   (b) Position indicator

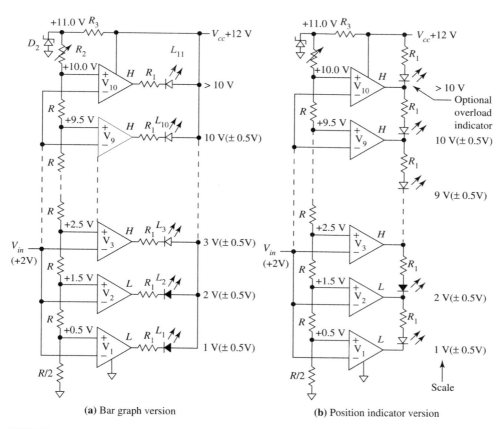

(a) Bar graph version   (b) Position indicator version

FIGURE 5–15   0 to 10-V bar graph and position type voltmeter.

The circuit consists of ten voltage comparators. The measured voltage is fed to the inverting inputs of the comparators. The non-inverting inputs are connected to a reference voltage string of 0.5, 1.5, 2.5, etc. volts. The comparators with reference voltages higher than the input voltage are set HIGH, the others LOW. The LED at the output of the comparator will light when the comparator is set LOW, generating a bar graph [Figure 5–15(a)]. The upper lit LED of the graph indicates the applied voltage with an accuracy of ±0.5 V. At the top of the graph is an LED that indicates overload when the applied voltage exceeds 10 V.

The position indicator version of the circuit is shown in Figure 5–15(b). Here, the LEDs are connected between the outputs of the comparators and will light only when they are between outputs of HIGH and LOW state. Therefore, only one LED will light.

The following are some design considerations. The resistor set R_1 determines the LED current and brightness. In the case of the bar graph version:

$$i_D = (V_{CC} - V_{LED} - V_{SAT})/R_1 \qquad \textbf{(5–5)}$$

where i_D = LED current (A)

$\quad V_{CC}$ = supply voltage (V)

$\quad V_{LED}$ = voltage drop in the LED (V)

$\quad V_{SAT}$ = saturation voltage of the comparator (V)

$\quad R_1$ = current limiting resistor (Ω).

In the case of the position indicator version, saturation voltages of the two comparators should be considered. The accuracy of the device is determined by the stability of the Zener diode voltage and the tolerances of the dividing string R. The transfer function of the comparator determines the precision of the switching point. Output and sinking current capability of the comparators should be considered when driving LEDs with high current.

In practice, one does not assemble bar graphs from components. Many integrated circuit chips that do the job in a much simpler way are available. A typical example is given in Figure 5–16, where two National LM3914 bar graph drivers are cascaded to form a 20-segment display. Only three external resistors are required: $R1$ and $R2$ determine the LED current and $R3$ bypasses the LED #10. An optional switch may be used to select between bar graph and position indicator modes. The chip has an internal reference supply of 1.25 V. A driver, National LM3915, with logarithmic response from 0 to 30 dB (decibels) in 3 dB steps is available. Chips are also available in integrated packages with an LED array and driver.

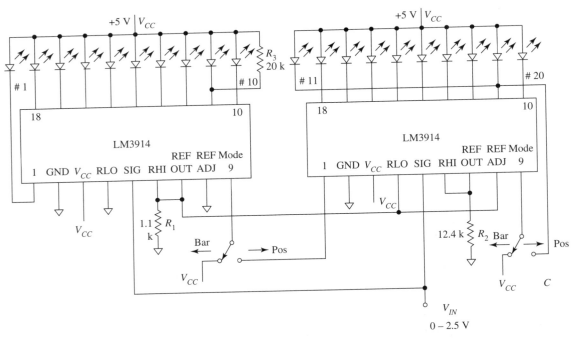

FIGURE 5–16 A bar or position graph using two cascaded LM3114 display drivers.

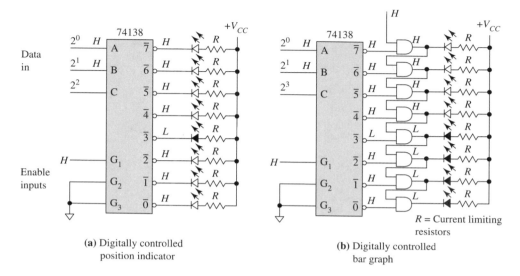

(a) Digitally controlled
position indicator

(b) Digitally controlled
bar graph

FIGURE 5–17   Digitally controlled bar graph or position indicator.

Besides bar graphs controlled by analog inputs, digitally controlled graphs are common. They are easy to design, as shown in Figure 5–17.

For a position indicator, a decoder or demultiplexer is needed. They are available for 1 of 4 to 1 of 16 outputs. An inverted output chip is usually preferred because the sinking capability of the chip is higher than the driving capability. Thus, most of the displays can be directly connected to the chip without a driver. In this case, the selected output is low, and the other outputs remain high. Therefore, only one LED connected to the low output will light [Figure 5–17(a)].

When a bar graph display is desired, an AND gate is connected to each output, as shown in Figure 5–17(b). As seen from the schematic, all AND gate outputs above the activated output stay HIGH and all outputs below, including the activated output, go LOW. Therefore all LEDs up to activated output, light.

Using two encoders and eight drivers, an 8-by-8 array can be addressed to make a 64-position indicator, as shown in Figure 5–18.

A six-bit word is used to control the array. The first three bits through the decoder IC1 set the selected column high. Depending on the required current output, drivers may be needed. The last three bits through the decoder IC2 set the selected row low. The LED at the junction of the selected column and row will light. With a proper multiplexing scheme, the array can be converted to a bar graph.

Readable displays are always a compromise between the quality or readability of the characters, commonly called *"font,"* and the complexity and cost of the drive circuit. The most economical are segmented displays.

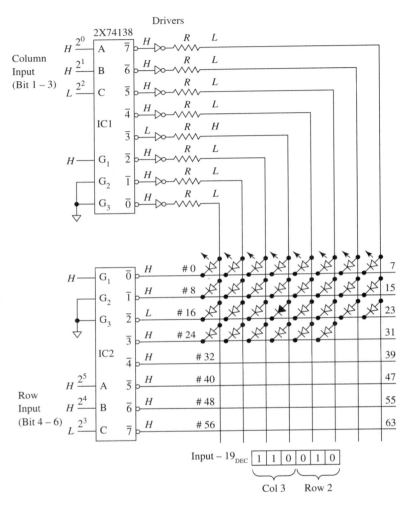

FIGURE 5–18 64-position indicator using two of 8 decoders.

The **segmented displays** are available in two configurations: **7-segment** and **16-segment displays,** as shown in Figure 5–19.

The 7-segment display is designed primarily for numerals and can display only a limited alphabet. The 16-segment display can present the entire upper case alphabet and numerals, as shown in Figure 5–20.

The set of 64 characters is addressed by a six-bit **ASCII code (American Standard Code for Information Interchange).** Because of the many interconnections required, the display is often an integrated unit of four characters requiring decoding and driving circuitry. The 16-segment font is not particularly legible and therefore has not found wide acceptance. In addition to the 16-segment display, a similar 14-segment display is also available.

FIGURE 5–19   16- and
7-segment displays.

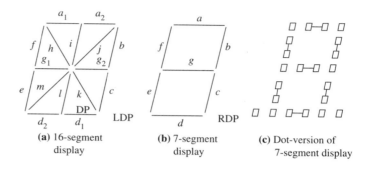

(a) 16-segment
display

(b) 7-segment
display

(c) Dot-version of
7-segment display

The 7-segment display is an industry standard for numeric displays. A typical display consists of seven bars with two decimal points, as shown in Figure 5–19(b). A dot version with 13 elements, as shown in Figure 5–19(c), is also available. The latter has a somewhat more legible font and allows display of a few more alphabetic characters. The fonts for both types are shown in Figure 5–21.

As we can see, a normal 7-segment display presents a fairly good numeric set and allows the presentation of 13 upper-case alphabetic characters. The lower-case characters are out of proportion and difficult to read. The quality of the dot version 7-segment font is considerably better and a few more upper-case letters can be presented, as seen in Figure 5–21(b). This display, however, requires a more complex and not so readily available decoder.

Decoders for 7-segment displays are available for all light sources, but only for numerals. Some manufacturers add a few upper-case letters to the decoder because the decoder requires a four-bit input code that allows the decoding of up to 16 characters or symbols. However, by programming a ROM, it is very easy to design a decoder, as demonstrated by Example 5–2.

EXAMPLE 5–2

A display is needed that produces the symbol set shown in Figure 5–22. Design a decoder for such a set.

BITS	D_3 D_2 D_1 D_0	0 0 0 0	0 0 0 1	0 0 1 0	0 0 1 1	0 1 0 0	0 1 0 1	0 1 1 0	0 1 1 1	1 0 0 0	1 0 0 1	1 0 1 0	1 0 1 1	1 1 0 0	1 1 0 1	1 1 1 0	1 1 1 1
D_6 D_5 D_4	HEX	0	1	2	3	4	5	6	7	8	9	A	B	C	D	E	F
0 1 0	2	(space)	!	"	∓	5	%	Ʀ	'	⟨	⟩	✳	+	,	−	.	/
0 1 1	3	0	1	2	3	4	5	6	7	8	9	:	;	∠	=	↘	?
1 0 0	4	@	A	B	C	D	E	F	G	H	I	J	K	L	M	N	O
1 0 1	5	P	Q	R	S	T	U	V	W	X	Y	Z	[	\	]	↗	⟨

FIGURE 5–20   16-segment ASCII character set. (Courtesy Hewlett-Packard, Inc.)

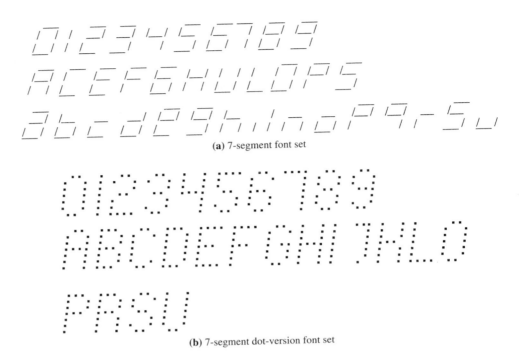

(a) 7-segment font set

(b) 7-segment dot-version font set

FIGURE 5–21   7-segment display fonts.

Solution

The easiest way to make a decoder is to program an eight-bit ROM so that each bit represents the status of each segment of the display. An Intersil IM5600 32 by 8 bits EPROM is sufficient for this application. The EPROM has an open collector output that can sink 16 mA, sufficient current to drive an LED. Therefore, a LOW output state is selected to activate a segment. The memory map for the EPROM is shown in Table 5–1. The circuit for the display is shown in Figure 5–23.

The LEDs, and most other displays also, can be divided into two groups: common anode and common cathode displays, as seen in Figure 5–24.

In a **common cathode** display, the segment is activated by applying voltage to the segment; in a **common anode,** by grounding the segment. The latter method has some advantages since most ICs have greater capacity to sink than to supply current. In many applications, this may eliminate the need for a driver.

For a display with few digits, three to four at most, the **direct drive** as shown in Figure 5–25 is the most cost effective.

Symbol number   0   1   2   3   4   5   6   7   8   9   10   11   12   13   14   15

Symbol

ROM Program for Experimental Symbol Set

DEC	HEX	A₄	A₃	A₂	A₁	A₀	O₇ DP	O₆ g	O₅ f	O₄ e	O₃ d	O₂ c	O₁ b	O₀ a	HEX	Symbol
		Binary														
0	00	0	0	0	0	0	1	1	1	1	0	1	1	1	F8	—
1	01	0	0	0	0	1	1	0	1	1	1	1	1	1	BF	—
2	02	0	0	0	1	0	1	1	1	1	1	1	1	0	FE	—
3	03	0	0	0	1	1	1	0	1	1	0	1	1	1	B7	=
4	04	0	0	1	0	0	1	0	1	1	1	1	1	0	BE	≡
5	05	0	0	1	0	1	1	0	1	1	0	1	1	0	B6	≡
6	06	0	0	1	1	0	1	1	0	0	1	1	1	1	GF	I
7	07	0	0	1	1	1	1	1	1	1	1	0	0	1	F9	I
8	08	0	1	0	0	0	1	1	0	0	1	0	0	1	C9	I I
9	09	0	1	0	0	1	1	1	1	1	0	0	1	1	F3	L
10	0A	0	1	0	1	0	1	1	0	1	1	1	1	0	DE	⌐
11	0B	0	1	0	1	1	1	1	1	1	1	1	0	0	FC	￢
12	0C	0	1	1	0	0	1	1	1	1	1	0	0	1	F9	⌟
13	0D	0	1	1	0	1	1	1	1	0	0	1	0	0	E4	⌐
14	0E	0	1	1	1	0	1	1	0	1	0	0	1	0	D2	⌐
15	0F	0	1	1	1	1	1	1	0	0	0	0	0	0	C0	□

FIGURE 5–22   Experimental symbol set.

FIGURE 5–23   Symbol encoder circuit.

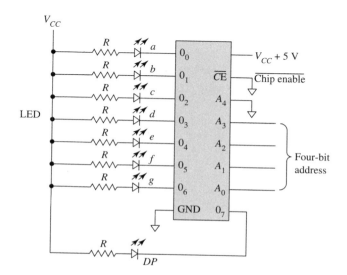

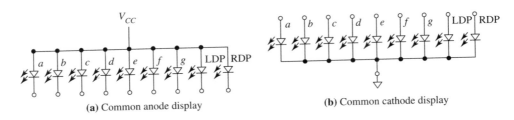

(a) Common anode display

(b) Common cathode display

FIGURE 5–24   Common anode and common cathode LED displays.

Drivers are needed when the display current or voltage requirement exceeds the capability of the encoder, or when special waveforms are needed, as in the case of LCD displays. In most cases, these are open collector transistor arrays. Often the drivers are part of the encoder, simplifing the circuit. A large selection of encoders and drivers is available.

The current through the segment is controlled with the resistor R. Its value can be calculated using Equation 5–4, considering the voltage drop in the segment, the saturation voltage of the driver, and the supply voltage:

$$i_s = (V_{CC} - V_S - V_D)/R \qquad \qquad \textbf{(5–6)}$$

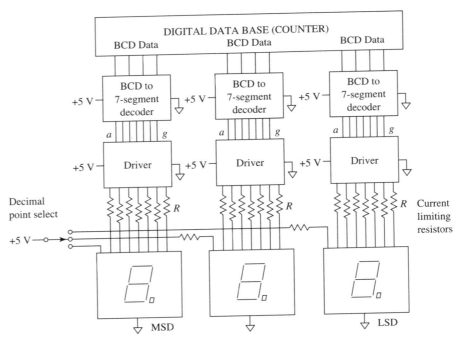

FIGURE 5–25   3-digit, direct drive cathode circuit.

where i_S = segment current (A)

V_{CC} = supply voltage (V)

V_S = voltage drop in the segment (V)

V_D = saturation voltage of the driver (V)

R = current limiting resistor (Ω).

With displays of more than four digits, a **strobed** or **multiplexed display** is more economical. In this system, only one decoder and driver are used. The output is shared by all digits using a multiplexer. The digits are energized only during a short period by using a heavy current pulse. In the case of LED displays, this arrangement also saves power since the LED is more efficient at high current, as described in Chapter 3. A principal strobed display circuit is shown in Figure 5–26.

The display requires a clock and a counter/divider to divide the clock frequency to the number of digits in the display (six in this case). The counter controls a multiplexer that, in turn, selects the digit to be activated at consecutive clock pulses. The BCD data are renewed and stored in six data latches. The multiplexer releases the data of the selected digit to the decoder, the driver, and the data bus connected to the segments of all digits. However, only the activated digit displays the data. With the next clock pulse, the next digit is activated and the data for that digit are selected, etc.

With N digits in the display, we can write:

$$t = t_P N \tag{5–7}$$

$$\text{and} \quad d_C = t_P/t \tag{5–8}$$

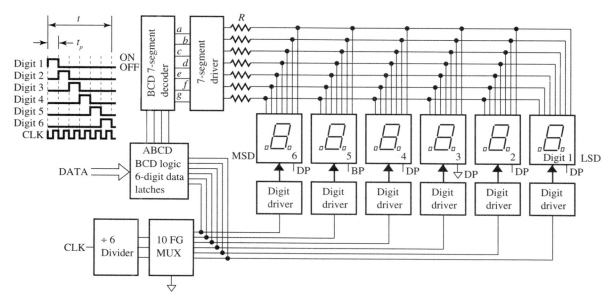

FIGURE 5–26   6-digit multiplex display.

where t_P = pulse time (s)
  t = refreshing period (s)
  N = number of digits in the display
  d_C = duty cycle of the applied waveform.

A number of design options should be taken into consideration. Clock frequency should be high enough to eliminate flicker. Since the human eye observes flicker at frequencies below 40 Hz, the longest refreshing period is $1/40 = 0.025$ s, or 25 ms. Therefore, the lowest clock frequency is:

$$f_{min} = N/0.025 \qquad\qquad \textbf{(5–9)}$$

where f_{min} = lowest flicker-free clock frequency (Hz).

The upper frequency limit is determined by the turn-off time of the bipolar drivers. At a strobing rate above 10 kHz, the digit driver may not be completely turned OFF and a "ghost" of the next frame may appear. This ghost may be eliminated by shortening the drive pulse to less than $1/N$ of the refreshing period.

The duty factor determines the peak current and luminous intensity of the display. It also establishes the current rating of the drivers. Example 5–3 demonstrates this relationship.

EXAMPLE 5–3

An HP 5082-7756 7-segment display is used in the circuit of Figure 5–26. The luminous intensity of the LED is 1.1 mcd at 20 mA, and the relative efficiency increases to 1.05 at 100 mA. The forward voltage drop is 1.7 V at 100 mA. The drivers have a peak current of 100 mA. Calculate the minimum clock frequency, value of the current limiting resistor R when a 5-V power supply is used, and the average luminous intensity of the display.

Solution
From Equation 5–7, the minimum clock frequency is:

$$f_{min} = N/0.025 = 6/0.025 = 240 \text{ Hz}$$

To simplify the filtering of the power supply, a somewhat higher frequency, 1 kHz for example, may be preferable.

The saturation voltage of a driver is about 0.9 V. Since two drivers are used, one for segments and one for the common cathodes, the current limiting resistor R can be calculated as:

$$R = (V_{CC} - V_{SEG} - 2V_{SAT})/i_{SEG} = (5.0 - 1.7 - 2 \times 0.9)/0.1 = 15\ \Omega$$

Using Equation 3–17, we can calculate the average luminous intensity as:

$$I_{avg} = I_{RO}I_P d_C \eta_P/I_O \eta_O$$
$$= 1.1 \times (100 \times 0.167 \times 1.05)/20 \times 1 = 0.96 \text{ mcd}$$

using the duty cycle of the waveform:

$$d_C = t_P/t = 1/6 = 0.167$$

The average luminous intensity is 0.96 mcd, somewhat less than the intensity at 20 mA steady current. The average current of the pulsed waveform, however, is:

$$i_{avg} = i_p d_C = 100 \times 0.167 = 16.7 \text{ mA}$$

which is also less than the 20 mA DC. Therefore, the efficiency of the pulsed current is somewhat higher than with DC.

5–5   DOT MATRIX AND LARGE-SCALE DISPLAYS

A 7-segment display is cost effective, but has two shortcomings. It can present only limited information with limited legibility. Even with a 16-segment display, the lower-case alphabet cannot be represented. A dot matrix display offers the simplest system that presents a full set of the lower- and upper-case alphabet and numerals at reasonable cost and complexity.

The most popular and simple dot matrix is a 5 × 7 array. It consists of 35 display elements set in a pattern of 5 columns and 7 rows. Each element can be addressed and energized by selecting the proper row and column, as shown in Figure 5–27.

This arrangement reduces the connections to the display from 35 to 12 but at the same time requires strobed operation of the display. Column strobing is shown in Figure 5–27 where energized rows for each column are shown. The arrows indicate selected rows and columns during each frame. Row-strobing can also be used.

The matrix allows a relatively good quality character set of upper- and lower-case alphabet, numerals, and symbols, as shown in Table 5–2.

The limited ASCII subset consists of 68 characters: upper-case letters, numerals, and some symbols; the complete set consists of 128 characters with lower-case letters, some foreign letters, and other symbols added. As one can see, quality and readability are considerably improved compared to a

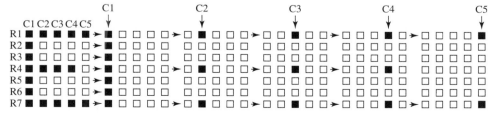

FIGURE 5–27   5 × 7 dot matrix column strobed display.

TABLE 5–2
5 × 7 Matrix ASCII Character Set (Courtesy Hewlett-Packard, Inc.)

		128 CHARACTER ASCII SET (HDSP-2471)							
			64 CHARACTER ASCII SUBSET (HDSP-2470)						
BITS D6 D5 D4 / D3 D2 D1 D0 COLUMN ROW		0 0 0 / 0	0 0 0 / 1	0 0 1 / 2	0 1 0 / 3	0 1 1 / 4	0 1 1 / 5	1 1 0 / 6	1 1 1 / 7
0000	0	NUL	DLE	SP					
0001	1	SOH	DC1						
0010	2	STX	DC2						
0011	3	ETX	DC3						
0100	4	EOT	DC4						
0101	5	ENQ	NAK						
0110	6	ACK	SYN						
0111	7	BEL	ETB						
1000	8	BS*+	CAN						
1001	9	HT*	EM						
1010	A	LF+*	SUB						
1011	B	VT	ESC						
1100	C	FF	FS						
1101	D	CR	GS						
1110	E	SO	RS						
1111	F	SI	US*						DEL*

*DISPLAY COMMANDS WHEN USED IN LEFT ENTRY
+DISPLAY COMMANDS WHEN USED IN RIGHT ENTRY

segmented display. For large-scale displays, CRTs, printers, etc., a complete ASCII set of 256 characters, as shown in Table 5–3, is available.

In the 7-segment display all segments can be simultaneously addressed during one address cycle. In a dot matrix display, each digit must be addressed either column by column or row by row. Therefore, each digit is made up of 5 or 7 subsets of data presented during a digit address cycle. Consequently, the display must operate as a strobed system, greatly reducing the interconnections and component count.

Input code to the display is a seven-bit word from the ASCII table that identifies the character. The heart of the system is a character-generator RAM that provides either a row or column data set for each character. Data

TABLE 5–3
Complete 256 Character Set

Dec.		0	16	32	48	64	80	96	112	128	144	160	176	192	208	224	240
	Hex.	0	1	2	3	4	5	6	7	8	9	A	B	C	D	E	F
0	0	∅	►	SP	0	@	P	`	p	Ç	É	á	░	└	╨	α	≡
1	1	☺	◄	!	1	A	Q	a	q	ü	æ	í	▓	┴	╤	ß	±
2	2	●	↕	"	2	B	R	b	r	é	Æ	ó	█	┬	╥	Γ	≥
3	3	♥	‼	#	3	C	S	c	s	â	ô	ú	│	├	└	π	≤
4	4	♦	¶	$	4	D	T	d	t	ä	ö	ñ	┤	─	┗	Σ	∫
5	5	♣	§	%	5	E	U	e	u	à	ò	Ñ	╡	┼	╞	σ	∫
6	6	♠	-	&	6	F	V	f	v	å	û	ª	╢	╞	╓	μ	÷
7	7	•	↨	'	7	G	W	g	w	ç	ù	º	╖	╫	╫	τ	≈
8	8	◘	↑	(	8	H	X	h	x	ê	ÿ	¿	╕	╚	╩	Φ	°
9	9	○	↓	)	9	I	Y	i	y	ë	Ö	⌐	╣	╔	╠	θ	∙
10	A	◙	→	*	:	J	Z	j	z	è	Ü	¬	║	╩	╟	Ω	·
11	B	♂	←	+	;	K	[	k	{	ï	¢	½	╗	╦	█	δ	√
12	C	♀	∟	,	<	L	\	l			î	£	¼	╝	╠	█	∞	ⁿ
13	D	♪	↔	–	=	M	]	m	}	ì	¥	¡	╜	═	█	ø	²
14	E	♫	▲	.	>	N	^	n	~	Ä	Pts	«	╛	╬	█	ε	■
15	F	☼	▼	/	?	O	_	o	⌂	Å	ƒ	»	┐	╧	█	∩	SP

latches, multiplexers, and frequency dividers are other components that coordinate the distribution of data. A principal three-digit 5 × 7 matrix schematic is shown in Figure 5–28.

Data for the display are stored in three-digit data latches. The clock releases these data to the character-generator ROM, where the sets of row data are generated. The clock selects the column and, at the same time, releases the appropriate row data to the row lines. Therefore, it takes five cycles per each digit to transfer all the data to the digits. When using LEDs, the maximum peak current for an element is about 100 mA. Considering the minimum average current of 1 mA, we can address only 100 columns, or a maximum of 20 digits. Therefore, the size of the display is limited.

This shortcoming can be overcome by adding storage latches to store the character-generator output and then load the data simultaneously to all digits.

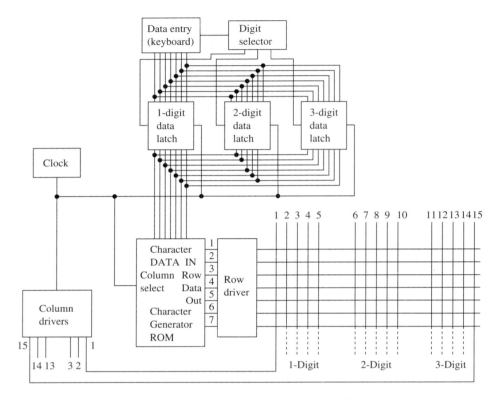

FIGURE 5–28   3-digit column strobed 5 × 7 dot matrix display.

A four-digit, row-strobed display using this idea is shown in Figure 5–29.

The input is similar to the previous schematic. The ASCII word for each digit is stored in digit data latches. Next, the timing circuit releases the first digit to the ROM that then decodes the first-row data and stores it in first-digit column latches. The same is repeated with all remaining digits. When this cycle is completed, all column latches are loaded with first-row data. The timing circuit then activates the first-row driver and the selected elements in all digits are activated simultaneously. The process repeats, decoding and loading the second row data into the column latches, and the clock activates the second row, etc. The advantage of this system is that the refreshing period consists of only seven cycles, independent of how many digits are addressed. Therefore, the duty cycle for each activated element is 1/7, and sufficient current to each element can be applied.

The drawback of this system is that it requires more hardware, including a latch and driver for each digit. However, there are somewhat more expensive display units where the latches and drivers are included in the display, eliminating the need for additional components.

Larger displays are often used in connection with a microprocessor. The microprocessor may be used as a data entry device and also fill the function of character generator. In most cases, these systems are designed by the manufacturers and are beyond the scope of this book.

Where low cost is a factor, the dot matrix display cannot compete with segmented displays. The latter can cost less than $1 per digit, whereas for dot matrix displays, the figure may approach $10 when all supporting circuitry is included.

When considering **large-scale displays,** the cathode ray tube (CRT) has no competitor. The CRT display has the lowest cost per character, is versatile, produces high quality fonts, has color capability, and is adaptable for high resolution graphics. However, it is bulky and requires a high-voltage

FIGURE 5–29   4-digit row-strobed 5 × 7 dot matrix display.

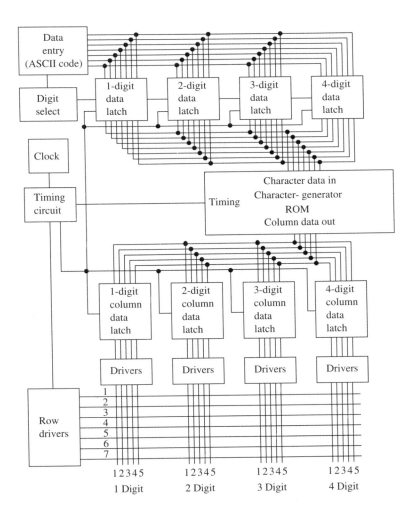

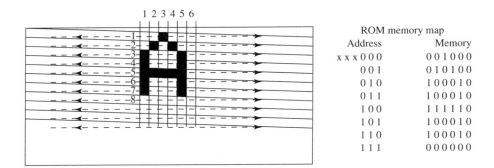

FIGURE 5–30   Scanned CRT display with character matrix.

power supply. Only where these shortcomings, especially the bulk, cannot be accommodated can large-scale flat panel displays compete with CRTs.

In practice, the average engineer seldom designs a CRT system. Such designing is left for the specialists in the field. Therefore, only the operating principles of this type of display are discussed below.

Unquestionably, the most popular CRT display system is the scanned matrix display. An electron beam scans the screen from left to right (horizontal scan) and steps the lines down (vertical scan) after each horizontal scan as shown in Figure 5–30. The characters are formed by turning the beam ON and OFF according to a digital pattern determined by the 6×8 character matrix. The pattern is similar to a 5×7 pattern except that it contains spaces between the characters and lines.

A principal schematic of a CRT drive is shown in Figure 5–31. The heart of the drive is a **Character Generator,** a $256 \times 8 \times 6$ ROM. The ROM contains the 6×8 matrixes of all ASCII characters and symbols. After the proper row is selected, the output of the matrix row is fed to a parallel-in/serial-out, 6-bit shift register. The output of this register is the signal that modulates the intensity of the beam, forming the elements, called **pixels,** of the video display.

The entire operation is synchronized by a quartz master clock that is digitally divided to generate operating cycles for the shift register, the horizontal and vertical syncronization, and a counter to select the matrix row. The resulting signals are shown in Figure 5–31. The complex circuitry to accomplish this is available in a single LSI chip.

A video RAM is used to store the ASCII character codes and addresses from the computer. It also controls the **character display attributes,** such as **intensified, reverse,** etc. For a color display, a color graphics adapter, basically a RAM that is programmed to drive three color guns, is used.

The quality of the display depends on the resolution, the number of pixels per line, and the number of lines available. See Table 5–4 for systems that are available.

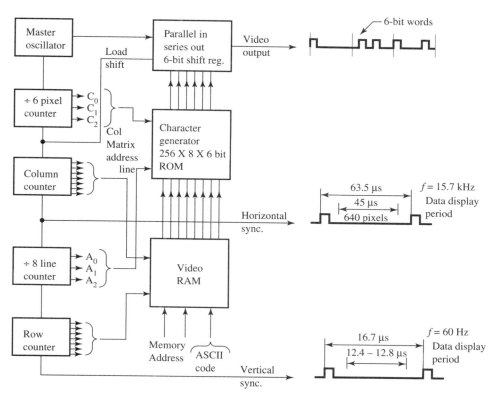

FIGURE 5–31   Principle schematic of CRT display drive.

TABLE 5–4
Available Display Systems

System	Horizontal Pixels	Vertical Lines	No. of Colors
CGA	640	200	Mono
RGB	640	200	16
EGA	640	350	16
VGA	640	480	16
VESA	800	600	256
8514	1024	769	Unlimited

SUMMARY

Chapter 5 discusses displays. It starts with the liquid crystal display, its operating principles, types of LCD displays, driving requirements, and relative advantages and limitations.

Requirements for visual displays—legibility, brightness, and contrast are introduced. Design of bar graphs and position indicators is described. The main emphasis of the chapter is on alphanumeric display, including 7-segment and 16-segment displays. Types of segmented displays and direct and multiplexed driving methods are described.

Operating principles of large-scale displays, such as dot matrix and CRT displays, are also introduced, along with a short description of driving methods and ASCII code and character sets.

RECOMMENDED READING

Bylander, E. G. 1979. *Electronic Displays*. New York: McGraw-Hill Book Company.

Chapell, A. 1978. *Optoelectronics: Theory and Practice*. New York: McGraw-Hill Book Company.

Hewlett-Packard. 1981. *Optoelectronics Fiber-Optics Applications Manual*. New York: McGraw-Hill Book Company.

Refliogu, H. I. 1983. *Electronic Displays*. New York: IEEE Press.

Bittleston, R. and Weston, G. F. 1982. *Alfanumeric Displays*. New York: McGraw-Hill Book Company.

PROBLEMS

1. An LCD is driven with waveform of 10% duty cycle and 5.00 V peak-to-peak value. Considering that the drive voltage should have zero average value, find the peak and RMS values for ON and OFF times.

2. Design the characters for an LCD reflective display that can be read from a 10 foot distance.

3. An HP 3900 series 7-segment display with luminous intensity of 3.3 mcd and segment area of 14.9 mm^2 is used as a display at daylight with 100 lx illumination. The background of the display has a reflection coefficient of 0.5. Find the contrast and contrast ratio and determine if the display meets the requirements shown in Figure 5–13. Do the same for bright daylight conditions with 1000 lx illumination.

4. Design a 5-lamp voltage indicator for a nominal Vs = 5.00-V power supply where lamp 1 (red) lights when Vs < 4.75 V; lamp 2 (yellow) when 4.95 < Vs < 4.75 V; lamp 3 (green) when 4.95 < Vs < 5.05 V; lamp 4 (yellow) when 5.25 < Vs < 5.05 V; and lamp 5 (red) when Vs > 5.25 V. Calculate the reference resistance divider for a position indicator. Use 6.30 V reference supply.

5. Design a 10-element bar graph VU meter which indicates from −4 dBm to +14 dBm in 2 dBm steps. The zero bB reference level in a 600 Ω is $V_0 = 0.773$ V, and the other voltages are calculated from $\Delta dB_1 = 20 \log(V_1/V_0)$. Use 10.00-V reference source.

6. A 7-segment LED display is driven directly with 30 mA. The voltage drop of LED is 2.15 V and dynamic resistance, 13.5 ohms. Find the current limiting resistance when a 5.00 V power supply is used.

7. An electroluminescent clock display has four digits plus a blinking second display with a steady-state luminance of 50 cd/m^2. Find the lowest clock frequency, duty cycle, and average luminance.

8. Find binary and hex column and row codes for the following characters in ASCII 128 character 5 × 7 matrix set: 4, K, Σ, r, Ü, and π.

9. Find primary and hex column and rpw code for the following characters and symbols from complete ASCII character set: 5, D, r, ä, ?, and Ω.

10. Design the letter E on a 6 × 8 matrix and write the memory map for the letter.

The answers are in Appendix C.

6
Radiation Detectors

The abundance of electromagnetic radiation that surrounds us is ultimately absorbed by receivers. All receivers respond to radiation in some way. A detector is a particular type of receiver exhibiting a response that can be used either to perform measurement, exercise control over other devices, communicate, or perform some other technical function. The great majority of radiation is absorbed by receivers that do not respond in this manner, but their importance should not be underestimated. Photosynthesis in plants and the warming of the earth's atmosphere are both the result of absorption by such receivers.

Detectors themselves also respond to radiation in different ways. In this book, we examine the detectors that have an electrical response since they are the easiest to apply to the useful purposes listed above. Unfortunately, this leaves out two of the most important detectors: the human eye, a biological detector, and photographic film, a chemical detector.

A great variety of detectors with electrical responses exist. Here we examine only those detectors that can be used as components or design elements by optoelectronics designers. More complex devices and their characteristics will only be listed.

6-1 CHARACTERISTICS AND CLASSIFICATION OF RADIATION DETECTORS

The **spectral response** describes the frequency or wavelength range over which a detector can respond. In this respect, detectors fall in two categories, wide or narrow spectral response, as depicted in Figure 6–1. Classifying

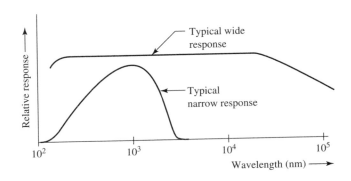

FIGURE 6–1 Typical wide and narrow response of detectors.

detectors by the operating principles, we can divide detectors into several groups. Two groups are most important.

Thermal detectors are devices that exhibit a wide response. They absorb radiation and react to the resulting temperature rise of the device. Therefore, their response depends only on the absorption characteristics of the device surface. With proper coating, the detector can absorb very long wavelengths in the long micrometer range. When detectors are enclosed in a glass envelope, the transmission characteristics of the glass modifies the response.

Photoelectric detectors are devices with a narrow response. In these devices, the detector response is caused by the direct interaction of photons with the atoms. The absorption of the photons depends on the available energy levels of the detector's atomic structure and is limited to relatively narrow spectral response.

Detector Characteristics

Many terms and parameters are needed to describe detector characteristics. The most significant are described below.

Responsivity (RE) describes the figure of merit of a detection system. It applies to detectors that respond to radiation by producing an electrical output, either current or voltage. The responsivity is the ratio of output to the radiant input:

$$RE = V_o/\phi_I \quad \text{or} \quad I_o/\phi_I \qquad (6\text{–}1)$$

where RE = responsivity (V/W), (V/lm), (A/W), or (A/lm)
V_o = output voltage from the detector (V)
I_o = output current from the detector (A)
ϕ_I = applied radiant or luminous flux (W), (lm).

The responsivity can be calculated either for a single wavelength or integrated over a given wavelength range (for black-body radiation, for example).

The **noise equivalent power (NEP)** is the power applied to the detector that produces an output signal equal to the RMS noise output from the detector. Since it is very difficult to detect radiation producing an output that is less than the inherent noise output from the detector, the *NEP* describes the lowest detectable radiation level of a given device. Besides responsivity, the *NEP* is also a figure of merit for a detector.

When noise output from a detector is known, the equation for *NEP* can be developed in the following way. When we use Equation 6–1, the signal output from a detector is

$$I_o = RE\,\phi_{\mathrm{I}}$$

According to the *NEP* definition, the applied power is equal to the *NEP* when the signal output equals the noise output I_{N}, or when the signal-to-noise ratio equals one:

$$S/N = 1 = I_o/I_N = RE \times NEP/I_{\mathrm{N}}$$

Therefore

$$NEP = I_{\mathrm{N}}/RE \qquad\qquad (6\text{–}2)$$

where NEP = equivalent noise power (W)
  RE = responsivity (A/W) or (V/W)
  I_{N} = RMS noise current or voltage output from the detector (A) or (V).

NEP is a useful parameter when comparing similar detectors under similar conditions. Under different conditions, the comparison does not apply. This is apparent from Equation 6–2, where both the factors I_{N} and RE depend on the external conditions. The noise output increases with the bandwidth, and RE decreases with increasing detector area (for a given radiant incidence, the incident flux to the detector increases with greater detector area, resulting in a decreasing RE). Therefore, a comparison of two detectors using *NEP* is valid only when they have the same area and respond to the same bandwidth.

A lower *NEP* designates a detector that is capable of more sensitive measurements. Since we are accustomed to using a higher number for a higher figure of merit, often a different term is used:

The **detectivity** of a detector is the reciprocal of *NEP*:

$$D = 1/NEP = RE/I_{\mathrm{N}} \qquad\qquad (6\text{–}3)$$

A more sensitive detector that can detect a lower level of radiation has a higher detectivity than a less sensitive one. Detectivity, as *NEP*, also depends on noise bandwidth and detector area. To eliminate these factors, a normalized figure of detectivity is used.

D* (Pronounced "dee-star") is detectivity normalized to a detector area of 1 cm² and a noise bandwidth of 1 Hz:

$$D^* = D\sqrt{A_{\mathrm{D}}\Delta f} = \sqrt{A_{\mathrm{D}}\Delta f}/NEP \qquad\qquad (6\text{–}4)$$

where D^* = normalized detectivity (cm $\times$ Hz$^{1/2}$/W)

$\quad A_D$ = detector area (cm^2)

$\quad \Delta f$ = noise bandwidth (Hz).

D^* can be used to objectively compare all detectors.

EXAMPLE 6-1

Find D^* of a detector with NEP of 12 pW, detector area of 10 mm^2, and noise bandwidth of 1 kHz. In the same system, find the improvement of NEP level when the noise bandwidth is reduced to 50 Hz.

Solution
For a 1-kHz system, using Equation 6–4,

$$D^* = \sqrt{A_D \Delta f}/NEP = \sqrt{(10 \times 10^{-2}) \times 10^3}/12 \times 10^{-12}$$

$$= 8.33 \times 10^{-11} \text{ cm Hz}^{1/2}/\text{W}$$

When the bandwidth in the same system is reduced to 50 Hz, the NEP value is

$$NEP = \sqrt{A_D \Delta f}/D^* = \sqrt{(10 \times 10^{-2}) \times 50}/8.33 \times 10^{-11}$$

$$= 2.68 \times 10^{-12} \text{ W} = 2.68 \text{ pW}$$

In the 50 Hz noise bandwidth system, NEP is reduced by the factor of 4.47.

Quantum efficiency describes the intrinsic efficiency of a photodetector. It is the ratio of the number of photoelectrons released to the number of incident quanta absorbed, at a given wavelength. An ideal photodetector with a quantum efficiency of 1 produces 1 electron per incident quanta. The quantum efficiency can be calculated from responsivity by the following equation:

$$\eta_\lambda = 1.24 \times 10^3 RE_\eta/\lambda \tag{6-5}$$

where η_λ = quantum efficiency at wavelength λ

$\quad RE_\eta$ = responsivity at wavelength λ (A/W)

$\quad \lambda$ = wavelength of the radiation (nm).

The quantum efficiency is a critical factor for designers of photosensitive devices but has less significance for the user.

Response time is a critical factor for many detectors, especially those used in communications. It can be expressed in two ways, as a time constant or as rise-and-fall time.

The time constant is used when the response is exponential, which is often the case with thermal detectors. It is the time the detector requires to reach $(1 - 1/e)$, or 63%, of its final response.

Rise-and-fall time is the time required to reach from 10% to 90% of the final response. Figure 6–2 explains these terms.

The rise time determines the highest signal frequency to which a detector can respond. A good approximation of the 3-dB point of the frequency response and rise time is expressed in the following equation:

$$f_{3dB} = 0.35/t_R \qquad\qquad (6\text{--}6)$$

where f_{3dB} = 3 dB point of the detector frequency response (Hz)
    t_R = rise time (s).

Detector Noise

Electricity (flow of electrons) and radiation (flow of photons) are both "granular" in their nature. The flow of elementary carriers in these phenomena is never quite steady but always exhibits small random variations. Thus, every signal carried by these media always contains a minute random frequency component, called **noise**.

Noise is the most critical factor in designing sensitive radiation detectors. The noise masks low-level signals and makes their recognition impossible. Thus, understanding the nature of noise and its origin is essential for low-level detection systems. Noise in these systems may be generated in radiation sources, detectors, and also post-detection devices and circuitry.

Johnson, Nyquist, or thermal noise is caused by the thermal motion of charged particles in a resistive element. This noise is generated in every resistor, regardless of its type or construction. The noise voltage or current depends on the resistance value, temperature, and the system bandwidth. It can be calculated as follows:

$$V_{Jrms} = \sqrt{4kRT\Delta f} \qquad\qquad (6\text{--}7)$$

$$I_{Jrms} = \sqrt{4kT\Delta f/R} \qquad\qquad (6\text{--}8)$$

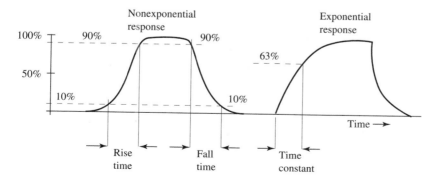

FIGURE 6–2   Definition of rise-and-fall time and time constant.

where V_{Jrms} = RMS noise voltage into load R (V_{rms})
   I_{Jrms} = RMS noise current into load R (A_{rms})
   R = resistance value (Ω)
   k = Boltzmann's constant 1.38×10^{-23} (J/K)
   T = absolute temperature (K)
   Δf = system bandwidth (Hz).

Shot noise, generated in a photon detector, is caused by the discrete or granular nature of the photoelectrons generated. It depends on the average current through a photodetector and the system bandwidth:

$$I_{Srms} = \sqrt{2eI_{avg}\Delta f} \tag{6–9}$$

where I_{Srms} = RMS shot noise current (A_{rms})
   I_{avg} = average current through the photodetector (A)
   e = charge of an electron 1.60×10^{-19} (C)
   Δf = system bandwidth (Hz).

Generation-recombination noise in a photoconductor is caused by the fluctuation in current carrier generation rate, recombination rates, or trapping rates in a photoconductor or semiconductor.

$$I_{GRrms} = 2eG\sqrt{\eta EA\,\Delta f} \tag{6–10}$$

where G = photoconductive gain where G = number of active electrons/
   number of photoelectrons generated
   η = quantum efficiency
   E = radiant incidence (W/cm^2)
   A = detector receiving area (cm^2).

This type of noise is predominant in photoconductive detectors operating at long (infrared) wavelengths.

Flicker or 1/f noise, occurs in all conductors where the conducting medium is not metal; for example, semiconductors and carbon. There is no good explanation for the origin of this noise. It depends on the semiconductor material used and its surface treatment. Also, an exact equation to calculate the noise does not exist, but it follows this relationship:

$$I_{Frms} = k\sqrt{I_{dc}^{a}\,\Delta f/f^{b}} \tag{6–11}$$

where I_{Frms} = RMS flicker noise current (A_{rms})
   I_{dc} = DC current through the conductor (A)
   Δf = system bandwidth (Hz)
   f = operating frequency (Hz)
   k, a, b = arbitrary constants.

The constant a is about 2, and constant b is about 1. The constant k depends on the semiconductor material and its treatment. The equation, which is not too useful for calculating the noise current, does indicate, however, one important characteristic of the noise: its amplitude is inversely proportional to the frequency. Usually this noise is predominant at frequencies below

100 Hz and exists in all semiconductor devices that require bias current for their operation. It also should be pointed out that, since the equation is empirical, the noise does not approach infinity at DC.

Noise is generated by a few other sources, including the microphonic noise due to vibration. In addition to noise in the detectors themselves, noise is also generated in amplifying devices that may follow the detector. Amplifiers have noise sources similar to those described above.

The total equivalent noise can be calculated by adding all noise voltages or currents in an *RMS* manner, as shown for the equivalent noise current equation below:

$$I_{Neq} = \sqrt{I_{Jrms}{}^2 + I_{Srms}{}^2 + I_{GRrms}{}^2 + I_{Frms}{}^2} \qquad (6\text{--}12)$$

where I_{Neq} = total equivalent *RMS* noise current (A_{rms}).

An example of calculating the effect of noise in a communication link is given in Chapter 8.

The most logical way to classify radiation detectors is by operating principle, shown in Figure 6–3. This drawing lists the most popular detectors. The emphasis of this text is on photoelectric detectors, because they are the most widely used and are easy to apply in optoelectronics designs. Modern detectors and the entire field of optoelectronics are based on semiconductor technology. There are other types of detectors, but they are too complex for use as simple design components and are only briefly described here.

Thermal Detectors

Thermal detectors, as the name implies, absorb radiation and operate on the resulting temperature rise. All types of thermal detectors have two common characteristics: (1) they depend on the heating of a mass, so their response time is slow; and (2) they can respond to a wide range of wavelengths since their response depends only on the absorption characteristics of that mass. Four types of devices are predominant: thermocouple, thermopile, bolometer, and pyrolytic detectors.

In **thermocouple type** detectors, the detecting element is a junction of two dissimilar metals. When the temperature of this junction changes, an electromotoric force (EMF) develops in the junction. The magnitude of the EMF, which is proportional to the temperature, depends on the metals used. Table 6–1 shows some common materials and the EMF developed:

TABLE 6–1
Commonly Used Thermocouple Materials

Materials	$\mu V/°C$
Bismuth–Antimony	100
Iron–Constantan	54
Copper–Constantan	39

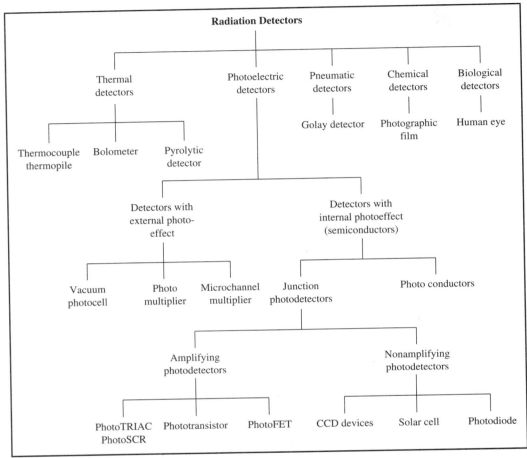

FIGURE 6–3 Types of radiation detectors.

An EMF also develops in every other junction of dissimilar metals in the circuit that is subjected to temperature change. These EMFs counteract the voltage in the measuring junction. Therefore, it is common practice to use two junctions, a measuring junction and a reference junction. The latter is kept at a constant temperature, often at 0°C in ice water. A typical thermocouple setup is shown in Figure 6–4.

The responsivity of a single thermocouple is very low and, therefore, it is mostly used as a thermometer since the thermal EMF as a function of temperature can be accurately calibrated. To increase the responsivity, several junctions can be connected in series to form a **thermopile**. Thermopiles are made either by connecting a series of wires or by using an evaporated thin film technique. The latter is constructed by depositing active bismuth and antimony junctions into a black energy-absorbing base, often mounted in a

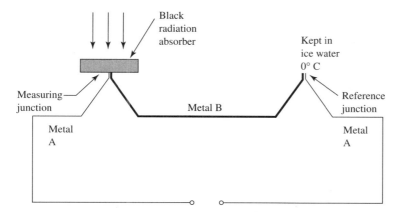

FIGURE 6–4   Thermocouple radiation detector.

hermetically enclosed envelope. The spectral response of the thermocouples extends into the far infrared to 40 μm. The frequency response of a thermopile is limited to approximately 20 Hz.

The **bolometer** is the most popular thermal detector. The sensing element in a bolometer is a resistor with a high temperature coefficient. This resistor is not a photoconductor. In a photoconductor, a direct photon-electron interaction causes a change in the conductivity of the material. In bolometers, the increased temperature and the temperature coefficient of the element cause the resistance change. The spectral response depends on the absorption characteristics of the element and is, therefore, a wide response, as with thermocouples.

A basic bolometer circuit is a bridge circuit, as shown in Figure 6–5. To reduce the effect of ambient temperature, usually two similar sensing elements are used. Only one element receives the radiation, but both are subjected to ambient temperature changes.

FIGURE 6–5   Basic bolometer circuit.

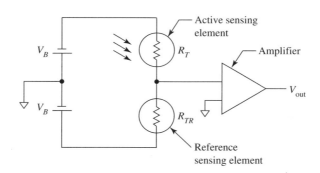

A **metal bolometer** uses a high temperature coefficient metal, such as bismuth, nickel, or platinum, with a temperature coefficient of about 0.3 to 0.5%/°C. This coefficient is relatively low and, therefore, the detectivity of the metal bolometer is also low. For a faster response time, a very thin film strip may be vacuum-evaporated to the absorbing base material.

The **thermistor bolometer** is the most popular. Because of its simplicity and relative sensitivity, it finds applications in burglar alarms, smoke detectors, and other similar devices. The detector in this bolometer is a thermistor, an element made of manganese, cobalt, and nickel oxides sintered together and applied to an insulating but thermally conducting wafer. The advantage of a thermistor is that it has a high temperature coefficient, up to 5%/°C, that varies with temperature as $1/T^2$.

More sensitive elements, typically $NEP = 10^{-10}$ W, have longer response times with time constants about 100 ms. Less sensitive detectors of $NEP = 10^{-8}$ W may have a time constant of 5 ms. The spectral response is typically 0.5 to 10 μm.

The **low-temperature germanium bolometer** is a sensitive laboratory type bolometer that uses a semiconductor (germanium) as the detector. The resistivity of gallium-doped germanium changes when radiated. The relative change of resistance is greatest, and therefore the responsivity highest, when the device is operated a few degrees above absolute zero temperature. Therefore, the device requires a cryogenic dewar to operate. The NEP of a germanium bolometer is 5×10^{-13} W, the time constant is fairly short at about 0.4 ms, the $D^* = 8 \times 10^{11}$, and the spectral response extends from 0.4 to 40 μm.

The **pyroelectric detector** is the most sensitive device that can operate without cryogenic cooling. The pyroelectric detector is based on the phenomenon that some dielectric materials exhibit a temperature-dependent polarization change. This results in a proportional dielectric constant change that is a function of temperature. Using such a dielectric, a capacitor can be built that changes its capacitance with temperature. It can be used as a sensitive temperature indicator or a thermal radiation detector.

The temperature/capacitance characteristics of a pyroelectric capacitor are shown in Figure 6–6. The device is operated at the point where the capacitor exhibits maximum responsivity; that is, where the capacitance change versus temperature is at a maximum, somewhere at the bend of the curve, at temperature T.

Pyroelectric material has another interesting trait: when heated above the temperature T_C, called the **Curie temperature,** the material loses its polarization and is no longer usable as a pyroelectric detector. The original characteristics can be restored by heating the material above the Curie temperature and applying a large voltage across the device. This process is called **poling**.

Pyroelectric materials include triglycerine sulfite, strontium barium niobate, lithium tantalate, and several others.

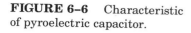

FIGURE 6–6   Characteristic of pyroelectric capacitor.

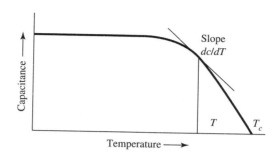

The application of a pyroelectric detector is simple, requiring a very high input impedance amplifier that can be operated either in a current or voltage mode. Typical circuits are shown in Figure 6–7.

The typical *NEP* of pyroelectric detectors is about 5×10^{-11} W, the *D** is about 10^8, the spectral response extends up to 50 μm, and response time is about 10 ms.

For **pneumatic detectors,** the **Golay cell** has the widest application. It is a small pneumatic chamber where gas is heated through a membrane. The expansion of the gas is detected either by electrical or optical devices. The cell operates at room temperature and has a very wide spectral range that reaches into the millimeter region. The response time is typically 2 to 30 ms, with $D^* = 10^9$.

The most widely used radiation detectors are the **human eye** and **photographic film.** They differ from other detectors in that they do not provide an output that can be used to control other devices, but they certainly can be used as measuring and communication devices having unique and interesting characteristics.

The human eye and photographic film, as used in a camera, respond to the luminance of the object instead of illuminance or sterance as the other detectors do. (The exception to this is photographic film used in X-ray

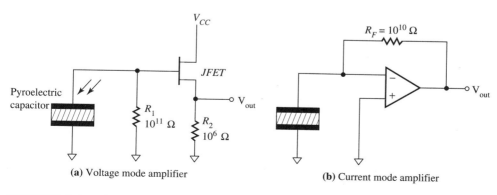

(a) Voltage mode amplifier   **(b)** Current mode amplifier

FIGURE 6–7   Application of pyroelectric detector.

technology, where the film is exposed without using the lens.) Second, both can give simultaneous information about the intensity and wavelength (or color) of the radiation, something no other detectors can do.

The characteristics of the human eye were described in Chapter 1. Its remarkable capability as a comparative measuring device deserves repeating. After all, the entire science of photometry was founded on two devices: the candela as an illumination intensity standard and the human eye as a detector. With clever instrumentation, fairly accurate and precise measurements are performed. The human eye still has its predominant place in optoelectronics.

Photographic film and photographic radiometry require great skill on sensitometry when quantitative measurements, even with modest accuracy and precision, are desired. The spectral range of film is from X-ray to about 1.2 μm. It requires 100 to 1000 photons to form a latent image on a silver halide emulsion. During the development, this minute exposure is developed to a measurable density on the photographic film—this is chemical amplification on an enormous scale. Also, keeping in mind that photographic exposure is an integrating process (that is, the photons can be collected over a long period of time), photographic film can detect radiation at the very lowest levels.

6–2 EXTERNAL PHOTOEFFECT PHOTOELECTRIC DETECTORS

Photoelectric detectors are devices in which electrons are released as the result of photon radiation. They can be classified into two groups: photodetectors with external and internal photoeffects.

In an **external photoeffect** detector the electrons are ejected from a photocathode surface, placed in vacuum, and collected by a positively charged anode. Vacuum photodiodes, photomultiplier tubes, and microchannel multipliers are typical examples.

Internal photoeffect takes place in semiconductors, of either junction or bulk type. The released electrons increase the conductivity of the detector or produce an amplified current output in active junction devices. Photoconductors, photodiodes, phototransistors, photo-FETs, photo-TRIACs, and photo-SCRs are typical examples.

Figure 6–8 shows the relative characteristics of both types of photoelectric detectors.

Vacuum Photodiode

Vacuum photodiode is the oldest electronic detector. It consists of a cathode and anode placed in a vacuum envelope, as seen in Figure 6–9.

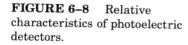

FIGURE 6–8   Relative characteristics of photoelectric detectors.

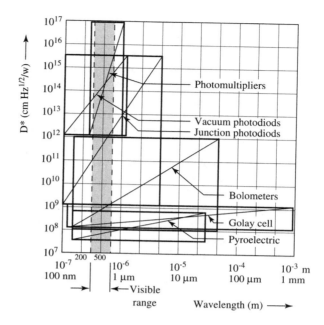

The operating principle of a vacuum photodiode is simple: the cathode, when radiated, releases electrons that are attracted by the positively charged anode. Thus, a photocurrent proportional to the photon flux results.

Besides vacuum photodiodes, gas-filled types are also available. Usually argon under low pressure (less than 1 Torr) is used. In these photodiodes, the electrons collide with gas atoms, causing ionization and increasing the current by a factor of about 10.

The quantum process in photodiodes is opposite to the process described for photoemitters in Chapter 3. The process for a metal cathode is shown in Figure 6–10.

When a photon with an energy level E_P strikes the photocathode, an electron is lifted to a higher energy level. When the photon energy is sufficient to overcome the cathode material energy gap ΔE, and survives internal collisions, the electron reaches a conduction band and is free to escape and generate a photocurrent. The energy gap, and therefore the spectral response

FIGURE 6–9   Vacuum photodiode.

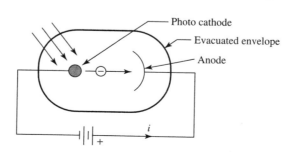

of the photodiode, depends on the cathode material. A higher energy band requires higher energy photons. This limits the spectral response of the metal cathode photodiodes to 300 nm, in the ultraviolet region.

Most photodiodes have semiconductor photocathodes. The energy gap in semiconductors is reduced by a process called **negative electron affinity, or NEA**, and extends their response toward infrared. Table 6–2 lists a few typical photocathode materials with their maximum response wavelength and quantum efficiency. The industry standard S designation, for different types of cathode materials, is also listed in the table.

The graph in Figure 6–11 shows the responsivity of these materials when used in a crown-glass vacuum envelope.

The outstanding characteristic of the photodiode is its very fast response time, on the order of 0.1 ns, and the wide linear range of the response that may extend over ten decades. Unfortunately, the lower part of this range is not usable because of noise, called **dark current.** The dark current may be calculated using the following equation:

$$i_T = aT^2A \exp(-eV_B/kT) = aT^2 A \exp(-1.16 \times 10^4 V_B/T) \qquad \textbf{(6–13)}$$

TABLE 6–2
Typical Photocathode Materials

Cathode Designation	Composition	Maximum Response (nm)	Quantum Efficiency (%)
S-1	AgOCs	800	0.4
S-10	BiAgOCs	420	6.8
S-11	Cs_3SbO	390	19.0
S-20	Na_2KSbCs	380	22.0
Bialkali	K_2CsSb	380	27.0

FIGURE 6–10 Atom energy level in photoemissive process.

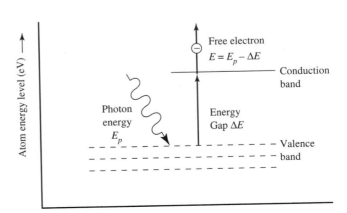

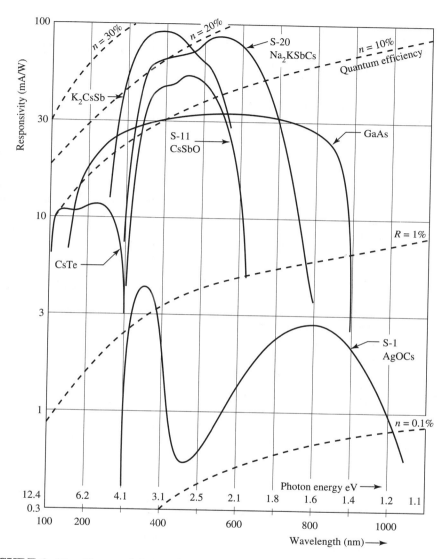

FIGURE 6–11   Responsivity and quantum efficiency of typical photodiode cathode materials.

where i_T = cathode dark current at temperature T (A)
    a = a constant depending on cathode material;
        for metals $a = 1.2 \times 10^6$ (Am^{-2} K^{-2})
    T = cathode temperature (K)
    A = cathode area (m^2)
    V_B = cathode barrier voltage (V)
    e = electron charge; $e = 1.60 \times 10^{-19}$ (C)
    k = Boltzmann's constant; $k = 1.38 \times 10^{-23}$ (J/K).

EXAMPLE 6–2

Calculate the dark current of a metal cathode photodiode with a barrier voltage of 1.25 V and an area of 100 mm^2 at 300 K.

Solution
Using Equation 6–13, we obtain

$$i_T = (1.2 \times 10^6)(10^{-14} \times 300^2) \exp[(-1.16 \times 10^4)(1.25/300)]$$
$$= 11.0 \times 10^{-15} \text{ A} = 11.0 \text{ fA}$$

As seen from Example 6–2, the dark current is so low that it is below the measuring range of the most sensitive current measuring device, the electrometer. Thus, the lower portion of the linear range of the vacuum photodiode is practically useless, and the limiting noise of the vacuum diode is the random noise of received radiation. This sets the *NEP* figure of the diode in the range of 10^{-14} to 10^{-16} W.

The vacuum photodiode requires relatively high anode voltage, usually in the range of 100 V, to operate, making it not directly compatible as a component in semiconductor circuitry. Also, the photodiode is somewhat bulkier than other photoelectric detectors.

The loss of the lowest operating range of the vacuum diode can be corrected with a photomultiplier.

Photomultiplier and Microchannel

The **photomultiplier** is a vacuum photodiode with a built-in low noise amplifier, as shown in Figure 6–12.

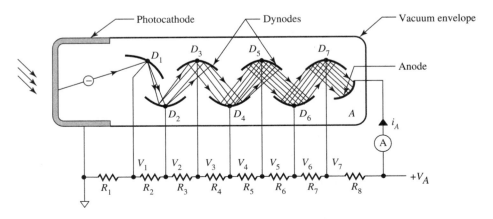

FIGURE 6–12   Photomultiplier.

The amplifier consists of a series of anodes, called **dynodes,** that are under progressively higher positive potential. The photoelectrons emitted from the cathode are accelerated toward the first dynode where they strike the dynode with an energy of eV_A, where V_A is the dynode voltage. Since the dynode work function is about 2 eV, the kinetic energy of the colliding electron releases more electrons from the dynode. Thus amplification takes place. The process is repeated to the next dynode until the amplified electron beam is collected by the final anode.

The multiplication of electrons, or the electron gain A, depends on the potential difference between the dynodes and the dynode material. In a metal dynode material, the striking electron will penetrate the surface and may be trapped. Therefore, an optimal acceleration voltage for maximum gain exists. In an **NEA-type dynode** the penetration and trapping of electrons is eliminated and the gain is a linear function of the accelerating voltage, as shown in Figure 6–13.

The total gain of the photomultiplier tube depends on the electron gain in each dynode and the number of dynodes:

$$G = A^n \tag{6–14}$$

where G = total photomultiplier gain
A = electron gain in a dynode
n = number of dynode stages.

EXAMPLE 6–3

Find total gain for a seven-stage B_e-C_u dynode photomultiplier with dynode voltage of 100 V.

FIGURE 6–13   Electron gain as function of dynode voltage.

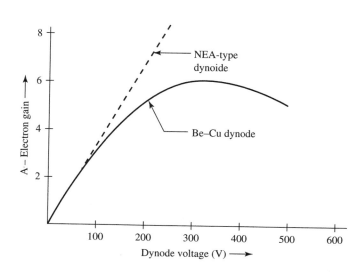

Solution

From Figure 6–13 for 100 V dynode voltage the electron gain is $A = 3.3$, the number of stages $n = 10$, and therefore the total gain is

$$G = A^n = 3.3^{10} = 153,200$$

All noise sources described in the previous chapter apply to photomultipliers. In addition, photomultipliers exhibit a specific noise, called **multiplication noise,** generated by the electron multiplication process in the dynodes. Despite this, and because of the built-in gain mechanism, the photomultiplier is a very sensitive, possibly the most sensitive, photodetector. It can detect the impact of a single photon!

The characteristics of a photomultiplier are as follows: the *NEP* is in the range of 10^{-16} (W), the D^* in the range of 10^{17} (cm Hz$^{1/2}$/W), and the rise time is about 2 ns, but the transit time from cathode to anode is on the order of 30 ns. The multiplier has a very wide linear range over eight decades. The spectral range is the same as that of a photodiode—from 100 to 1000 nm.

Besides the focused dynode version shown in Figure 6–12, other versions of photomultipliers are available. They may have a circular configuration, venetian blind-type dynodes, dynodes with control grids, and other configurations. They all require a well-regulated, stable high-voltage power supply and are relatively expensive. Therefore, they cannot be considered as components but rather, apparatuses. A few typical photomultipliers are shown in Figure 6–14.

The **microchannel** operates on the same principle as the photomultiplier—the direct amplification of photoelectron flux. It is illustrated in Figure 6–15.

The electron flux amplification in a microchannel takes place in a glass tube. The inside of the tube is coated with a mixture of oxides of silicon, lead, and alkali compounds. This coating has the property of providing an electron gain of two, for each wall collision. With two electrodes, high voltage, in kilovolts, is applied to the channel, accelerating the electrons from the

FIGURE 6–14 Typical photomultipliers. (Courtesy of Hamamatsu Corporation.)

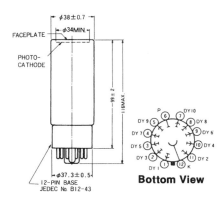

FIGURE 6–15 Micro-channel photodetector.

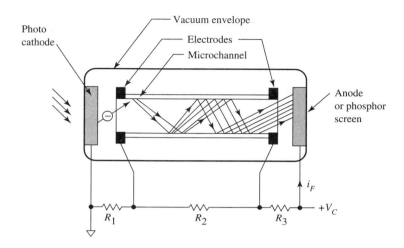

photocathode through the channel. At the output end of the channel, the electrons are collected by an anode or directed to a phosphor screen.

The microchannel is used in two modes: as a sensitive radiation detector, similar to a photomultiplier, or as an amplifying imaging device.

In a photodetector application, a single channel several centimeters in length and up to 10 mm in diameter is used. The channel may provide an electron gain approaching 10^8 and linearity over five decades. It does not equal the performance of a photomultiplier, but is a much simpler device and can operate in the presence of magnetic fields.

For an image amplification application, a plate with a tightly drilled hole pattern (15 μm in diameter) is used. A high voltage (about 1 kV) is applied to the surfaces of the plate. The holes act as a microchannel and amplify the optical image received from the photocathode and project it to the phosphor screen where an amplified image can be seen. Most night vision devices use this type of microchannel plate.

6–3   INTERNAL PHOTOEFFECT PHOTOELECTRIC DETECTORS

In internal photoeffect detectors, as in external ones, the photons release electrons. The electrons, however, remain in the photosensitive medium, changing its conductivity or generating current within it. Without exception, internal photoeffect detectors are semiconductor devices. They have become the building blocks of modern optoelectronics technology because they are inexpensive, reliable, small, and very versatile.

Internal photoeffect detectors are classified as either photoconductors or junction devices. **Photoconductors** are bulk semiconductor materials that change conductivity when radiated. **Junction devices** have one or more *p-n* junctions and act as diodes or transistors whose operation is controlled by radiation.

Photoconductors

Photoconductors are semiconductors that release electrons when radiated. As a result, the conductivity of the semiconductor increases. The quantum mechanism of the photoconductor is the same as that of a photodiode, as shown in Figure 6–10. The absorbed photon must have enough energy to elevate the electron across the energy gap to the conduction band, where it is free to move and carry a charge. Moving an electron from the valence band generates a hole, which can also act as a carrier. If an electric field is applied, the electron and hole move in opposite directions.

The spectral response of the photoconductor is determined by the energy gap of the semiconductor material used for the detector. Since the photon energy is proportional to the radiation frequency (or inversely proportional to the wavelength), the maximum absorption wavelength for a semiconductor can be calculated from the energy gap:

$$\lambda_{max} = 1240/\Delta E \qquad (6\text{--}15)$$

where λ_{max} = longest wavelength a semiconductor can absorb (nm)
ΔE = semiconductor energy gap (eV).

Typical photoconductor materials with their energy gaps and maximum wavelengths are given in Table 6–3.

Other intrinsic photoconductor materials are doped germanium and silicon semiconductors. They have an energy gap as low as 0.04 eV or a maximum response wavelength of over 30 μm. Because of their low energy gap, they also exhibit strong thermal current and require cryogenic cooling for efficient operation.

TABLE 6–3
Characteristics of Photoconductive Materials

Material	Energy Gap (eV)	λ_{max} (nm)
PbSe	0.23	5390
PbS	0.42	2590
Ge	0.67	1850
Si	1.12	1110
CdSe	1.80	690
CdS	2.40	520

Below the maximum wavelength, the responsivity of the photoconductor decreases linearly with the wavelength, as shown in Figure 6–16.

The D^* value of photoconductors listed above is about 10^8 to 10^9. Intrinsic or doped photoconductors operating at cryogenic temperatures reach D^* values up to 10^{11}. Therefore, they compare very favorably with thermal detectors in applications where wide spectral response is not required and simplicity and low cost are paramount.

The conductance of a photoconductor increases (or the resistance decreases) with increasing irradiance or illuminance. The relationship is close to a logarithmic function over a four- to five-decade range, as shown in Figure 6–17.

The resistance-illuminance relationship can be described with the following equations:

$$R_a = R_b(E_a/E_b)^{-\alpha} \tag{6-16}$$

or

$$E_a = E_b(R_a/R_b)^{-1/\alpha} \tag{6-17}$$

where R_a = resistance at illumination E_a (Ω)
  R_b = resistance at illumination E_b (Ω)
E_a, E_b = illumination (lx, fc)
  α = characteristic slope of the resistance-illumination curve.

The slope α between points a and b of the resistance-illuminance curve can be calculated as follows:

$$\alpha = \tan\theta = |(\log R_a - \log R_b)/(\log E_a - \log E_b)|$$
$$= |\log(R_a/R_b)/\log(E_a/E_b)| \tag{6-18}$$

The slope of the resistance curve is typically 0.55 to 0.9.

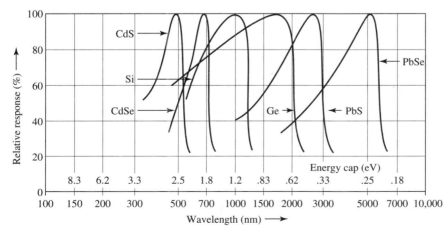

FIGURE 6–16   Spectral response of photoconductor materials.

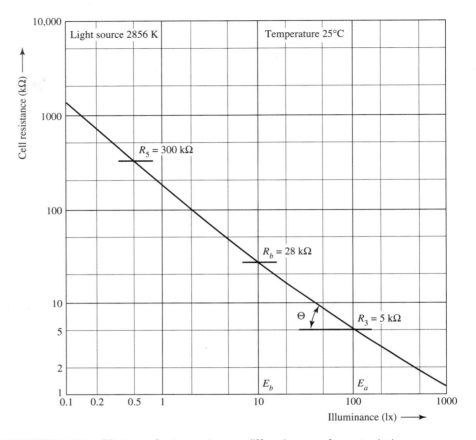

FIGURE 6–17   Photoconductor resistance/illuminance characteristic.

EXAMPLE 6–4

From Figure 6–17 calculate the slope between illumination of 10 lx and 100 lx and a resistance value of the photoconductor at 0.5 lx.

Solution
From Figure 6–17:

$$E_a = 100 \text{ lx} \qquad R_a = 5.00 \text{ k}\Omega$$

$$E_b = 10 \text{ lx} \qquad R_b = 28.0 \text{ k}\Omega$$

Using Equation 6–18, we obtain

$$\alpha = |\log(R_a/R_g/\log(E_a/E_b)|$$

$$= |\log(5.00/28.0)/\log(100/10.0)| = 0.748$$

From Equation 6–15, we obtain

$$R_{0.5} = R_b(E_{0.5}/E_b)^{-\alpha}$$

$$= 28.0 \times (0.5/10)^{-0.748} = 263 \text{ k}\Omega$$

As seen from Figure 6–17 the actual resistance at 0.5 lx is 300 kΩ. This discrepancy is typical because α varies slightly along the resistance-illuminance curve.

Photoconductor data sheets give either α and resistance values at some given illumination or present a set of curves. Information is also given about the illumination sources to which the curves are matched. Industry practice presents the curves for a tungsten source with a color temperature of 2856 K, that is, the color temperature of an ordinary light bulb.

The resistance value and the responsivity of the detector depend on the photosensitive material and the construction of the detector. Since most devices used in connection with detectors do not respond to resistance change, but require voltage or current change, the photoconductor requires a bias source and a load resistor R_L to operate, as shown in Figures 6–18(a) and (b).

The responsivity of other detectors is measured by their output current or voltage divided by the applied flux or illumination (A/ϕ or V/ϕ). Photoconductors, however, change resistance value when radiated so a different term for responsivity is used. For photoconductors, the responsivity is expressed by the relative change of resistance divided by the relative excitation change, either of flux or illumination:

$$RE_R = (\Delta R/R)/(\Delta E/E) \tag{6–19}$$

where RE_R = resistance responsivity
$\Delta R/R$ = relative resistance change
$\Delta E/E$ = relative excitation change, either flux or illumination.

According to this definition, when a 10% illumination change causes a 4% resistance change, the responsivity is 0.4. Using Equations 6–16 and 6–17, we can develop:

$$1 - \Delta R/R = (1 + \Delta E/E)^{-\alpha}$$

and for small values of $\Delta R/R$ and $\Delta E/E$, an approximation:

$$RE_R = (\Delta R/R)/(\Delta E/E) = -\alpha \tag{6–20}$$

or

$$\Delta R/R = -\alpha \Delta E/E \tag{6–21}$$

Thus, for a photoconductor with $\alpha = 0.5$, a 10% illumination increase causes a 5% resistance decrease, and with $\alpha = 1.0$, a 10% illumination increase causes a 10% resistance decrease.

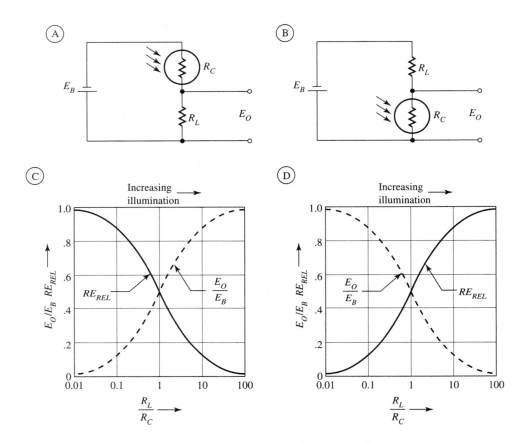

FIGURE 6–18   Photoconductor bias circuit and corresponding response curves.

Using a photoconductor in a bias circuit as shown in Figure 6–18 decreases the responsivity of the conductor. The relative change of the circuit resistance is smaller because of the load resistance R_L. This effect is depicted in Figures 6–18(c) and (d), where the ratio RE_{RC}/RE_R is the factor by which the natural responsivity of the photoconductor is decreased. The choice of R_C and R_L also affects the output voltage from the circuit. Thus the designer has a choice between higher output voltage or responsivity, as shown in Figure 6–18.

We can define the relative resistance responsivity as $(\Delta R_c/R_c)/(\Delta\phi/\phi)$. The responsivity is zero when the cell resistance R_c does not change when radiated and is 1 when $\alpha = 1$. Using the photoconductor in a biased circuit requires a series resistor R_L, changing the responsivity since relative resistance change is now $\Delta R_c/(R_c + R_L)$. Increasing the load resistor R_L increases output voltage from the circuit, but at the same time decreases the responsivity of the photoconductor. This condition is depicted in Figures 6–18(c) and (d) where the relative responsivity RE_{rel} is compared to the unbiased photo-conductor, together with the relative voltage output as shown.

EXAMPLE 6–5

Design a circuit that turns a light ON when illumination falls below 15 fc and analyze the precision of the switching circuit. Available components are photoconductor R = 5.00 kΩ at 100 lx, α = 0.8, 5.00-V power supply, 3.00-V reference diode, and a voltage comparator.

Solution

The circuit is shown in Figure 6–19. The light is activated when the voltage VB_o exceeds 3.00 V.

   The illumination level 15 fc corresponds to E = 15.0 × 10.76 = 171.4 lx. At that level, the photoconductor resistance may be calculated by using Equation 6–16:

$$R_a = R_b(E_a/E_b)^{-0.8} = 5.00(161.4/100.0)^{-0.8} = 3.41 \text{ k}\Omega$$

In this condition, V_{RC}/V_{RL} = 2.00/3.00 or

$$R_L = R_C(3.00/2.00) = 3.41 \times 1.50 = 5.11 \text{ k}\Omega$$

   The responsivity of the photoconductor is RE_R = −0.8. The circuit responsivity is reduced by a factor as shown in Figure 6–18. For R_L/R_C = 1.5, the factor is about 0.4. Thus, the responsivity of V_o is 0.8 × 0.4 = 0.32. A 10% illumination decrease causes the output voltage to change by 3.2% or 96 mV. The comparator switches on at about 2 mV differential. Thus, the switching precision is 2/96 of 10% or 0.2% of the illumination change. Later discussions show that this high resolution is masked by other characteristics of the photoconductor.

   For a 5-mA current through the zener diode, the value of the resistor R_1 is

$$R_1 = (V_{cc} - V_z)/i_z = (5.00 - 3.00)/5.00 = 0.40 \text{ k}\Omega$$

FIGURE 6–19   Photo-conductor light-activated switch.

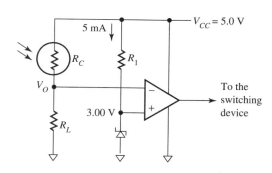

The construction of the photoconductor is influenced by the fundamental laws of photoconductor operation, which can be described with the following equations:

$$RE_I = (q\lambda\eta/hc)G \tag{6-22}$$

where G = photoconductive gain (free electrons/photon).

The photoconductive gain depends on the type of semiconductor used and the geometry of the device:

$$G = \tau_L \mu E/l = \tau_L \mu V/l^2 \tag{6-23}$$

where τ_L = free carrier lifetime (s)
μ = mobility of the carriers (cm/Vs)
E = electric field strength (V/cm)
l = length of the channel (cm).

It is evident from this equation that to achieve maximum gain, the photoconductive channel must be short and wide, as shown in Figure 6–20.

The photoconductive cell is constructed on a ceramic base onto which a photoconductive layer is evaporated or, in case of a sintered cell, painted and fused. On top of the conductive layer two electrodes are formed, leaving a narrow gap for the photoconductive channel. Leads are fastened to the electrodes, and the entire assembly is either coated in plastic or hermetically sealed for protection against humidity.

As with every electronic component, a photoconductive cell has ratings for maximum dissipation and applied voltage, as shown in Figure 6–21.

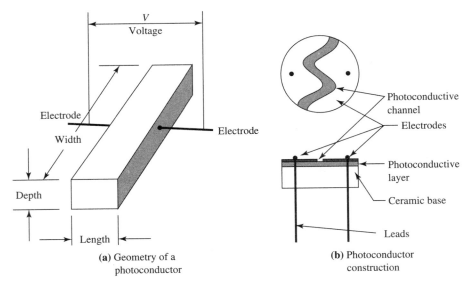

(a) Geometry of a photoconductor

(b) Photoconductor construction

FIGURE 6–20   Configuration of a photoconductor.

FIGURE 6–21
Photoconductor cell maximum
dissipation and voltage limits.

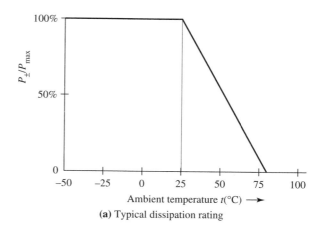

(a) Typical dissipation rating

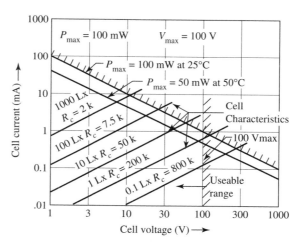

(b) Dissipation and max voltage
limits in I_c/V_c chart

Besides the dissipation and maximum voltage limits, cell tolerances should be considered. Typically, as specified in the data sheet, cell resistance has a very wide tolerance of ± 30 to $\pm 50\%$. Therefore, designs should always be checked for extreme limits, or the components should be selected with narrower tolerances.

Very wide tolerances are also typical of other photoresistor characteristics. Rise and decay time constants depend not only on semiconductor material, but vary widely with illumination level, as shown in Figure 6–22.

As we see from the graphs, the rise and decay time constants can be very long, on the order of 1 second at low illumination levels. At best, the time constants are in the range of a few milliseconds, making this particular detector unsuitable for the communication field that requires fast data rates.

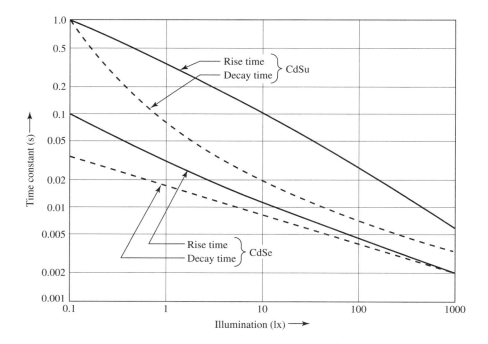

FIGURE 6-22   Typical photoconductor rise and decay time constants.

The millisecond range is, however, quite satisfactory for control, card reading, and similar applications.

Photoconductors do not rate highest in temperature stability. They exhibit a high temperature constant that is also illumination-dependent, as shown in Figure 6-23.

The graphs show that temperature characteristics vary greatly with semiconductor material, manufacturing method, and illumination level, so a fixed temperature coefficient for this type of detector cannot be specified.

Photoconductors exhibit a very specific phenomenon not found in any other detector: the **memory effect.** At a given illumination the cell may

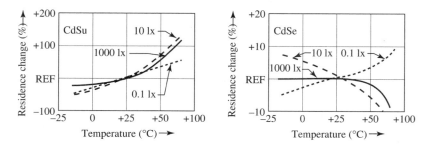

FIGURE 6-23   Typical temperature effect of photoconductors.

have several resistance values, depending on prior cell condition. This phenomenon is illustrated Figure 6–24, where the cell resistance is measured at 1 lx after either being illuminated for a long time (several hours) with 1000 lx or kept in total darkness. After being exposed to 1000 lx, the cell has a resistance value R_L that is higher than the value R_D, which is measured after the cell is kept in darkness. The ratio R_L/R_D may have a value of 1.1 at 1000-lx illumination level or up to 5 at the 0.1-lx illumination level. Therefore, the effect is significant. It will take minutes or hours at 1-lx illumination before the final and stable value of the resistance is reached.

It should be mentioned that the graphs presented here are for illustration purposes to indicate the magnitudes of the effects. They should not be used for design purposes. Consult the manufacturer's data sheet for a specific device.

The photoconductor is not stable and is not suitable for precise and accurate measurement. However, it is an inexpensive, simple, and durable device that has many applications in light switching, dimmers, punch-card readers, photoelectric relays, flame monitors, etc. For reliable operation, however, the wide variations of its characteristics should always be considered, as shown in Example 6–6.

EXAMPLE 6–6

Design an LED brightness control circuit that drives 10 mA through the LED at 100 lx illumination, and 20 mA at 500 lx. Use a photoconductor with $R_C = 5.00$ kΩ at 100 lx and $\alpha = 0.8$.

Solution
A simple transistor drive as shown in Figure 6–25 can be used. The task is to drive the base circuit so that it causes emitter voltage to change in a 2-to-1 ratio when illumination changes from 500 lx to 100 lx.

FIGURE 6–24   Memory effect in photoconductors.

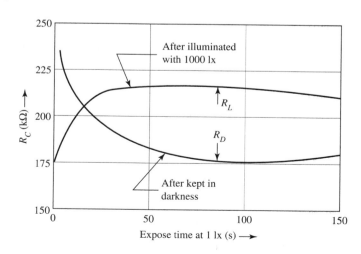

FIGURE 6–25   LED
brightness control.

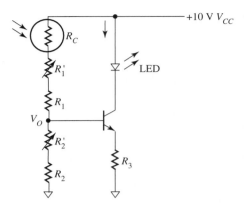

From Equation 6–16, we get

$$R_{C500} = R_{C100}(100/500)^{-0.8} = 1.38 \text{ k}\Omega$$

Selecting $V_E = 2.00$ V at 500 lx, we can write

E	V_E	$V_B = V_E + 0.7$	R_C
(1x)	(V)	(V)	(kΩ)
500	2.00	2.70	1.38
100	1.00	1.70	5.00

The current ratio through the base string should be

$$i_{500}/i_{100} = V_{B500}/V_{B100} = (R_{C500} + R_S)/(R_{C100} + R_S) = 2.70/1.70$$

where $R_S = R_1 + R_1' + R_2 + R_2'$.

From here we can write

$$1.70(R_{C500} + R_S) = 2.70(R_{C100} + R_S) \quad \text{and} \quad R_S = 4.76 \text{ k}\Omega$$

The current through the base string at 500 lx is

$$i_{500} = V_{CC}/(R_{C500} + 4.76) = 10.0/(1.38 + 4.76) = 1.63 \text{ mA}$$

Thus,

$$R_2 + R2_2' = V_{B500}/i_{500} = 2.70/1.66 = 1.66 \text{ k}\Omega$$

and

$$R_1 + R_1' = R_S - (R_2 + R_2') = 4.76 - 1.63 = 3.10 \text{ k}\Omega$$

Since the tolerance of photoconductors may be ±50%, the fixed value of R_1 and R_2 is selected at 50% at the calculated value and a variable resistor at the calculated value is added. This allows a change in resistor value by ±50% and compensates for photoconductor tolerances.

Junction Photodiode

The junction photodiode is a *p*- and *n*-type semiconductor junction similar to the junction used in an LED. The function of the photodiode junction, however, is the opposite of an LED junction. In the latter, photons are released in response to the current flow through the junction. In a photodiode, the photons are absorbed, resulting in the generation of free carriers that manifest as current through the junction. The principal operation of the photodiode junction is illustrated in Figure 6–26.

In an *n*-type semiconductor, electrons are the free carriers. Because of the excess electrons, the average energy level is higher than that of a *p*-type semiconductor. When a *p-n* junction is formed, the average energy level of both sides of the junction must be the same and, therefore, the valence and conduction band energy levels on both sides of the junction are different. The difference can be expressed as voltage V_o, as shown in Figure 6–26(b).

At the *p-n* junction, an excess of electrons on the n side and an excess of holes on the *p* side exist. Since the electrons and holes are mobile, they can combine and neutralize one another and, therefore, form a region on both sides of the junction where there are no carriers. This area, called the **depletion region,** acts as an insulator since it has no carriers. The diffusion of carriers from the p- and n-sides to the region is blocked because, after the free carriers are neutralized, ionized atoms of opposite polarity remain. This action generates a **barrier voltage** that opposes further diffusion of the free

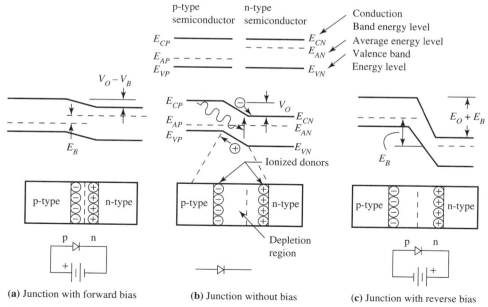

(a) Junction with forward bias   (b) Junction without bias   (c) Junction with reverse bias

FIGURE 6–26   Energy levels and depletion regions in a *p-n* photodiode.

carriers into the junction. The barrier voltage depends on the semiconductor material. For silicon it is $V_o = 0.7$ V. The junction therefore looks like a capacitor, a nonconducting dielectric separating two conductors.

When forward bias is applied to the junction, as in Figure 6–26(a), it opposes the barrier voltage, reducing the width of the depletion region and increasing the junction capacitance. When the bias reaches the barrier voltage, the depletion region is eliminated and the junction becomes conductive.

When reverse bias is applied, the depletion region is widened, the junction capacitance reduced, and the junction stays nonconductive, as seen in Figure 6–26(c).

The reverse bias junction, however, can conduct current when free carriers are introduced into the junction, either by thermal excitation or by radiation. How the diode behaves under these conditions is best analyzed using the equivalent circuit of the junction, as shown in Figure 6–27.

The junction may be regarded as an ideal diode, where the resistance of the depletion region is represented as R_{SH} and the junction capacitance as C_D, both connected parallel to the diode. The bulk resistance of the n- and p-side semiconductor of the junction is represented as a series resistor R_S. The current introduced to the junction by radiation is represented by a constant current source i_λ, parallel to the diode. For a silicon photodiode, the typical values are:

$$R_{SH} = 10^7 \text{ to } 10^{12} \text{ ohms, depending on temperature}$$
$$C_D = \text{tens of picofarads, depending on reverse bias}$$
$$R_S = \text{few ohms.}$$

Typical photodiode characteristics are shown by curves in the dark and radiated conditions in Figure 6–28.

The characteristic curve has four quadrants. In quadrant I, the diode is forward biased and acts in a way similar to a regular junction diode, a condition that is not suitable for a photodiode application. In quadrant II, the diode has no response. Quadrant III shows the diode's characteristics in a **reverse bias mode,** the predominant operation mode for radiation detection.

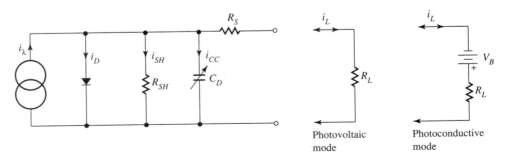

FIGURE 6–27 Photodiode equivalent circuit.

FIGURE 6–28   Junction photodiode characteristics.

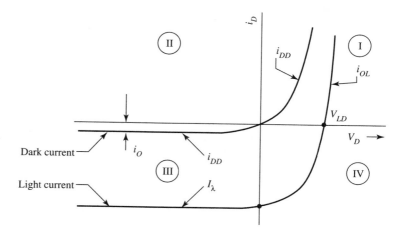

Quadrant IV is the **photovoltaic** or **solar cell mode** where the diode is used to provide power.

Kirchof's law applied to the diode equivalent circuit yields

$$i_\lambda = i_D + i_{SH} + i_L \tag{6–24}$$

where i_λ = photocurrent at wavelength λ (A)
  i_D = current through the ideal diode (A)
  i_{SH} = current through the shunt resistance (A)
  i_L = current into the load (A).

The characteristic curves of the photodiode can be developed from this relationship. For nonirradiated (dark) conditions the current is

$$
\begin{aligned}
i_{DD} &= i_o[\exp(eV_D/kT) - 1] \\
       &= i_o[\exp(1.16 \times 10^4 \times V_D/T) - 1]
\end{aligned} \tag{6–25}
$$

where i_{DD} = diode current without radiation (A)
  i_0 = diode dark current (A)
  V_D = diode voltage (V)
  e = electron charge; $e = 1.60 \times 10^{-19}$ (C)
  k = Boltzmann's constant; k $= 1.38 \times 10^{-23}$ (J/K)
  T = absolute temperature (K).

The dark current depends on the semiconductor material, doping, junction geometry, and ambient temperature. The dark current increases by a factor of 10 for a 25°C increase of temperature. Along with diode noise, the dark current limits diode sensitivity. Dark current at 25°C is usually given in photodiode data sheets.

When irradiated, a photocurrent i_λ is generated. Its value is

$$i_\lambda = \eta IAe\lambda/hc = 8.04 \times 10^5 \times \eta IA\lambda \tag{6–26}$$

where i_λ = photocurrent at wavelength λ (A)

$\quad\quad\eta$ = quantum efficiency (electrons/proton)

$\quad\quad I$ = irradiance (W/cm^2)

$\quad\quad A$ = diode area (cm^2)

$\quad\quad$h = Planck's constant; h = 6.62×10^{-34} (Js)

$\quad\quad$c = speed of light; c = 3.00×10^8 (m/s).

The photocurrent is a linear function of irradiance, holding over a range of seven to nine decades. This property makes the photodiode a valuable measurement device in the reversed bias or photoconductive mode.

The light current curve follows a function:

$$i_{\mathrm{DL}} = i_o[\exp(eV_{\mathrm{D}}/kT) - 1] - i_\lambda$$
$$= i_o[\exp(1.16 \times 10^4 \times V_{\mathrm{D}}/T) - 1] - i_\lambda \quad\quad\quad \textbf{(6–27)}$$

In an open-circuit condition ($R_{\mathrm{L}} >> R_{\mathrm{SH}}$), the photodiode develops the open circuit voltage

$$V_{\mathrm{LO}} = kT/e \, \ln[(i_\lambda + i_o)/i_o] \quad\quad\quad \textbf{(6–28)}$$

Under normal conditions $i_\lambda >> i_o$, and the equation simplifies to

$$V_{\mathrm{LO}} = (kT/e)\ln(i_\lambda/i_o) = 0.862 \times 10^{-4} \, \mathrm{T} \, \ln(i_\lambda/i_o) \quad\quad\quad \textbf{(6–29)}$$

At constant temperature, the open circuit voltage from a photodiode is a logarithmic function of irradiance, with a range of several decades. The device is useful for measuring where the logarithmic response is desired; photographic light meters and densitometers are typical examples of this application. It should be pointed out, however, that the condition $R_{\mathrm{L}} >> R_{\mathrm{SH}}$ is not easy to meet since R_{SH} is in the range of hundreds of megaohms.

In quadrant IV, a photodiode can deliver power into a load, as in solar (photovoltaic) cells.

Junction photodiodes can be classified into two groups: the solar (photovoltaic) cells and low-power photodiodes. The latter are called signal photodiodes because their main application is signal detection for communication, control, or metrology. This group includes many types, with the planar, PIN, avalanche, and Schottky diodes being the most widely used.

The purpose of **photovoltaic** or **solar cells** is to efficiently convert radiant energy to electric energy. High quantum efficiency is therefore paramount, and it is important to match the spectral response of the detector to the spectral energy distribution of the source (solar radiation, in this case). As seen in Figure 6–29(a), a semiconductor photodiode can capture only a small fraction of the solar emission spectrum. To improve the efficiency, a sandwich-type construction is used. The lower layers are designed to capture the radiation that is outside the response range of the top layer [see Figure 6–29(b)]. Typical single-cell efficiency is about 12%, and the efficiency of a triple-layer cell approaches 18 to 20%.

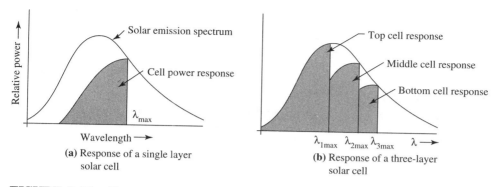

(a) Response of a single layer
solar cell

(b) Response of a three-layer
solar cell

FIGURE 6–29   Response of single and multi-layer solar cell.

To get maximum power from a solar cell, matching the load resistance to cell characteristics is most important. This process is illustrated in Figure 6–30, which depicts the characteristics of a Vactec type VTS28 that has a cell area of 0.6 in.^{2} and a short-circuit current of 86 mA at a solar irradiation of 100 mW/cm^2 (full sunlight).

Three load lines, with load resistances of 4.5, 5.7, and 7.5 ohms, are drawn on the graph. They represent the following output conditions:

R_L (Ω)	V_O (V)	i_O (mA)	P_O (mW)
4.5	3.5	78	27.3
5.7	4.0	70	28.0
7.5	4.5	60	27.0

The load line that forms the biggest area, defined by the crossing point of the characteristic curve and i/V axes, produces maximum output power. As

FIGURE 6–30   Load line
selection for solar cell.

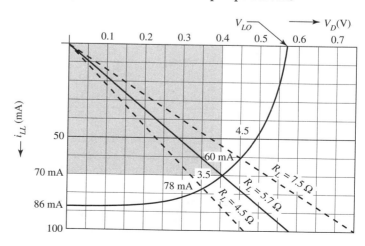

seen from the example, the output power curve is quite flat and therefore the selection of the load resistor is not critical. If the characteristic curve of the cell is not given, selecting a value $R_L = R_S$ gives a satisfactory result:

$$R_{\text{Lopt}} = R_S = V_{\text{LO}}/i_L \qquad \qquad \textbf{(6–30)}$$

In this case the value of $R_L = 0.57/0.086 = 6.6$ ohms.

In practice, solar cell output seldom matches the load requirement of the device to be powered, so a solar array with several parallel and series cells is used. The number of serial and parallel cells can be calculated:

$$n_P = \text{int}(i_D/i_L) \quad \text{and}$$

$$n_S = \text{int}(V_D/V_L) \qquad \qquad \textbf{(6–31)}$$

where n_P = number of parallel cells
   n_S = number of serial cells
   V_D = required device voltage (V)
   i_D = required device current (A)
   i_L = cell output current (A)
   V_L = cell output voltage (V)
   int = next higher integral of the quotient.

EXAMPLE 6–7

Design a solar array to charge a lead storage battery with 200 mA at 2.3 V. Use solar cells with characteristics as shown in Figure 6–30.

Solution

$$V_D = 2.3 \text{ V}; \qquad V_o = 0.4 \text{ V}$$

therefore

$$n_S = 2.3/0.4 = 5.75 \rightarrow 6 \text{ cells}$$

$$i_D = 200 \text{ mA}; \, i_L = 70 \text{ mA}$$

therefore

$$n_P = 200/70 = 2.86 \rightarrow 3 \text{ cells}.$$

The array is shown in Figure 6–31.

The spectral response of solar cells is typically from 400 to 1050 nm with a peak at 875 nm. The responsivity at peak sensitivity is about 0.5 A/W. Since solar cells are designed for high irradiance operation, the noise consideration and D^* term do not apply.

Signal photodiodes are junction diodes designed to operate as accurate and stable measuring and communication devices. The emphasis of their design is on a wide linearity range, fast response, and high thermal stability. Modern devices have achieved excellence in all these areas.

FIGURE 6–31   Solar array.

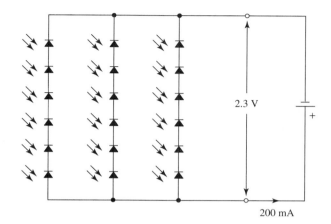

Signal photodiodes are mostly used in quadrant III with reversed bias for linear operation. When a logarithmic response is desired, they must be operated in quadrant IV in an open circuit mode, as shown in Figure 6–32.

The reverse bias can have any value below the maximum breakdown voltage V_{Rmax} that is specified in the data sheet. Increasing bias voltage increases the dark current and linear operating range and decreases the response time and junction capacitance. Typical relationships are shown in Figure 6–33.

A photodiode is usually used in conjunction with an operational amplifier. Two of the most common circuits are shown in Figure 6–34.

FIGURE 6–32   Photodiode reverse bias operation.

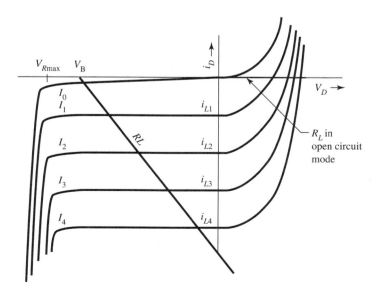

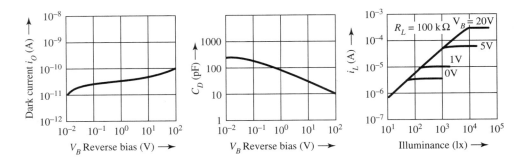

FIGURE 6–33   Effect of bias voltage on signal diode.

It should be noted that in circuit (b), the photodiode with resistance R_F forms the feedback loop of the amplifier. Because the diode is not pure resistance, but is shunted with capacitance C_D, the frequency response of the amplifier may peak or the amplifier may even oscillate at higher frequencies. To eliminate this problem, shunting of the feedback resistor R_F with a capacitor C_F may be needed. When use of the entire linear range of the photodiode (seven to nine decades) is desired, special consideration should be given to diode and amplifier noise. When fast response time is desired, as in high communication rate links, diodes with low C_D and low load resistance should be selected.

EXAMPLE 6–8

Design a three-range illumination meter with 100, 1000, and 10,000 lx full scale.

Solution

To make a true illumination measurement, the photodiode spectral response should be close to the C.I.E. standard observer curve. Hamamatsu S133 with

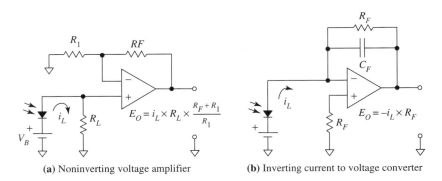

(a) Noninverting voltage amplifier   (b) Inverting current to voltage converter

FIGURE 6–34   Principle operational amplifier circuits for photodiodes.

FIGURE 6–35   3-range illumination meter circuit.

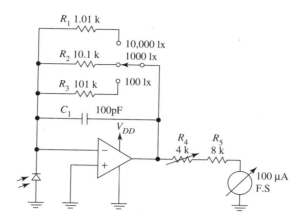

responsivity of 0.3 A/W for 3000 K color temperature source and with an area of 6.6 mm^2 has such response. From Figure 1–13 (pg. 22) we find that the efficacy for the 3000 K source is 20 lx/W. Thus diode photometric responsivity is

$$RE_P = 0.3 \text{ A/W} = 0.3/20 \text{ A/lm} = 0.015 \text{ A/lm}$$

Illuminance of 10,000 lx to the area of 6.6 mm^2 corresponds to

$$\phi_{10,000} = EA = 10^4 \times 6.6 \times 10^{-6} = 6.6 \times 10^{-2} \text{ lm}$$

Diode photocurrent at 10,000 lx is

$$i_{10,000} = RE_P\phi = 0.015 \times 6.6 \times 10^{-2} = 0.99 \times 10^{-3} = 0.99 \text{ mA}$$

Correspondingly,

$$\text{at 1000 lx} \quad i_{1000} = 0.099 \text{ mA} = 99 \ \mu\text{A}$$

and

$$\text{at 100 lx} \quad i_{100} = 9.9 \ \mu\text{A}$$

Selecting a 100 μA full-scale meter with 10 kΩ series resistor for indicator, the full-scale indication requires a 1.00 V output voltage from the operational amplifier. From this we can compute the feedback resistors R_1 to R_3:

$$\text{10,000 lx range} \quad R_1 = V_o/i_{10,000} = 1.00/0.99 = 1.01 \text{ k}\Omega$$

$$\text{1,000 lx range} \quad R_2 = 1.00/0.099 = 10.1 \text{ k}\Omega$$

$$\text{100 lx range} \quad R_3 = 1.00/0.0099 = 101 \text{ k}\Omega$$

The circuit is shown in Figure 6–35 where the needed offset adjustments are omitted. To allow ±20% adjustment for possible diode characteristics, the 10 kΩ series resistor is replaced with an 8.00 kΩ fixed resistor and a 4 kΩ variable resistor. A 100 pF capacitor parallel to the feedback resistors is

needed to compensate diode parallel capacitance to reduce possible peaking
or oscillation of the circuit.

When a logarithmic response from the diode is required, the diode
should operate without bias in an open circuit condition. Since the diode
shunt resistance is on the order of 10^8 ohms, the amplifier used should have
an extremely high input resistance. This effect can be achieved using FET
devices, teflon terminals, and guard circuits, as shown in Figure 6–36.

EXAMPLE 6–9

Using the same photodiode as in Example 6–8, design a photographic
incident-light exposure meter with logarithmic scale and full-scale indication of 10,000 lx. Calculate the indication at 1/10 of full scale.

Solution
Two solutions are possible: use the photodiode in the linear mode followed by
a logarithmic amplifier or use the photodiode in the logarithmic (open circuit)
mode with a linear amplifier. We select the latter, more economical, solution.

To drive a photodiode with internal resistance of 10^8 ohms requires a
very high input resistance operational amplifier. A FET input operational
amplifier (National LF155) has an input resistance of 10^{12} Ω and is well
suited for this purpose.

The open circuit voltage according to Equation 6–29 is

$$V_{OL} = 0.862 \times 10^{-4}\ T\ \ln(i_\lambda/i_o)$$

where

ambient temperature	$T = 290$ K
signal current from Example 6–8 at 10,000 lx	$i_\lambda = 0.99$ mA
dark current from data sheet	$i_o = 10$ pA.

Thus, the open circuit voltage is

$$V_{OL} = 0.862 \times 10^{-4} \times 290 \times \ln(0.99 \times 10^{-3}/10 \times 10^{-12})$$

$$= 0.862 \times 10^{-4} \times 290 \times 18.4 = 0.460\ \text{V}$$

FIGURE 6–36   Photodiode
response for logarithmic
response.

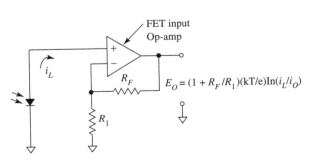

FET input
Op-amp

$E_O = (1 + R_F/R_1)(kT/e)\ln(i_L/i_O)$

Selecting, again, an indicator that requires 1.00 V for full-scale indication, the gain A of the amplifier is

$$A = V_o/V_{in} = 1.00/0.460 = 2.17$$

From the circuit (Figure 6–37) selecting $R_1 = 1$ MΩ we derive

$$A = (R_1 + R_2)/R_1 \quad \text{or} \quad R_2 = AR_1 - R_1 = 2.17 \times 1.00 - 1.00 = 1.17 \text{ MΩ}$$

Calibration to diode tolerances is achieved making indicator series resistance variable—8 kΩ fixed and 4 kΩ variable.

At full scale, the term $\ln(i_\lambda/i_0)$ has a value of 18.4. At 1/10 of full scale, the value is 1.84. Thus, the diode current is:

$$\ln(i_\lambda/i_0) = 1.84 \quad \text{or} \quad i_{\lambda 1/10} = i_0 \exp 1.84 = 10 \times 10^{-12} \times 6.30 = 63.0 \text{ pA}$$

Since the diode current is proportional to the illumination, the illumination at 1/10 scale is

$$E_{1/10} = E_{FS}(i_{1/10}/i_{FS}) = 10^4(63.0 \times 10^{-12}/0.99 \times 10^{-3}) = 0.636 \text{ mlm}$$

The range of the meter scale is therefore

$$E_{FS}/E_{1/10} = 10 \times 10^4/0.636 \times 10^{-3} = 1.57 \times 10^8$$

which corresponds to 27 zones (a zone is $2 \div 1$ illumination range, or 27 zones is a range of $2^{27} = 1.34 \times 10^8 \approx 1.57 \times 10^8$). A normal photographic illumination meter range has about 13 zones. Thus, our meter may not have sufficient resolution and the scale should be expanded using standard methods. The offset adjustment circuits are not shown on the schematic.

Junction photodiode spectral response and characteristics are determined by the design of the diode and the semiconductor material used. Most common types are described in the paragraphs that follow.

The **planar diffusion diode** is the basic junction-type photodiode. Its construction is shown in Figure 6–38.

At the top of the diode is heavily doped p-type material embedded in a lightly doped n-type material. At the junction of the two materials, a depletion region forms. Because of the different doping levels, the depletion level

FIGURE 6–37 Photographic incident light meter circuit.

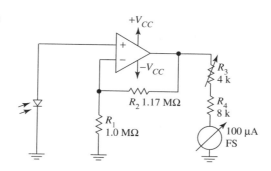

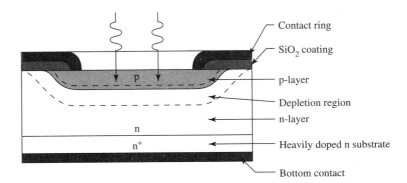

FIGURE 6–38   Planar junction photodiode.

reaches deeper into the n-type material than into the p-type material. At the bottom of the diode there is usually a layer of n^+-type material, called **substrate,** providing a connection with the bottom metal contact. The top contact is fused to the semiconductor over a transparent SiO_2 layer.

The penetration of photons through the layers depends on the wavelength of radiation. Shorter wavelengths (ultraviolets) are absorbed at the surface, while the longer wavelengths (infrareds) can penetrate deep into the structure. A wide response photodiode should therefore have a thin p-layer and a thick depletion layer so that most of the protons will be absorbed in the depletion layer.

The response time of the diode is controlled by the diode capacitance which, in turn, is determined by the thickness of the depletion layer and by the reverse bias. The thickness of the layer can be controlled by the doping level of the n-layer. Lower doping levels increase the thickness of the depletion layer and reduce the diode capacitance. This technique is used in the **PIN type photodiode,** where a thick layer of low doped and highly resistive n-type material is inserted between the p- and n-layers. This middle layer is called the **intrinsic** or **I-layer;** hence the name **PIN diode.**

The PIN diode has a relatively thick depletion layer which, with a modest reverse bias of 5 V, can be expanded to the bottom of the intrinsic layer. The result is a photodiode with a faster response time and an extended spectral response.

The **Schottky photodiode** is a junction diode where the p-n junction is replaced with a metal semiconductor junction, as shown in Figure 6–39.

A thin transparent layer (less than 10 nm) of gold is sputtered on the top of the n-type semiconductor (usually GaAsP or GaP). A junction forms between the gold and the semiconductor. Since the depletion layer of the junction is very thin, a detector with an enhanced UV response can be designed.

The **avalanche photodiode** is a junction diode with an internal gain mechanism. The device works only in a reverse bias condition. The electrons generated by the photons are injected into the conduction band. Due to the

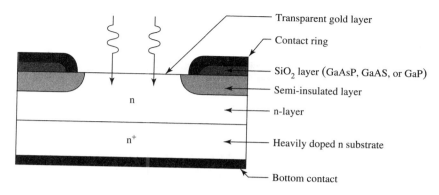

FIGURE 6–39   Schottky photodiode.

high electric field, the electrons accelerate to a velocity where, in collision with other atoms, they can release more carriers (electrons and holes). Therefore, a multiplication process, or amplification, takes place. Since the process is random it also generates shot noise, but under some conditions the S/N ratio of the diode can be improved.

The current gain depends on the reverse bias and the junction geometry. A typical relationship is shown in Figure 6–40.

The gain, as seen from the graph, is highly dependent on the bias voltage, and for stable operation the bias must be stabilized. The maximum gain is limited where thermally generated carriers are amplified. Therefore, the avalanche effect must be limited to the area of the junction where photoelectrons are released. This effect is achieved by surrounding the exposed junction area with a low doping level guard ring. In the guard ring the depletion range is much wider, and the field strength lower, so the current gain is smaller. An avalanche diode is depicted in Figure 6–41.

FIGURE 6–40   Current gain/bias relationship in an avalanche photodiode.

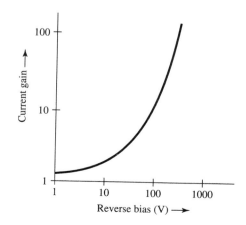

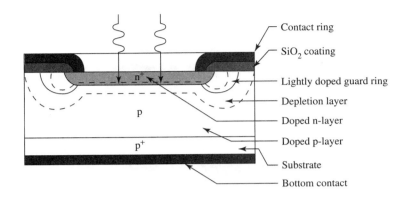

FIGURE 6–41   Avalanche photodiode.

Avalanche diodes are often used in communication and fiber-optics applications where the need for fast rise time dictates the use of low load resistance, resulting in low voltage responsivity. The internal gain mechanism of avalanche diodes has a considerable advantage for this application.

Compared to photoconductors, junction diodes are very stable devices. Their temperature coefficients are wavelength-dependent, but at maximum response wavelength the coefficients are very low, less than 0.1%/°C. The same can be said of aging effects. Therefore, the junction photodiode has many applications in metrology where high accuracy, stability, and a wide linear range are required. The only temperature-sensitive characteristics of the diode are dark current and shunt resistance, but their effects are significant only at very low level operation. A typical relationship of these characteristics is depicted in Figure 6–42.

The spectral response of junction photodiodes ranges from ultraviolet to infrared, depending mainly on the semiconductor material, but also on the junction design and window material. Examples of typical spectral responses

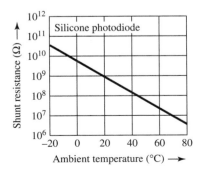
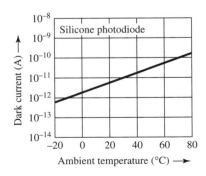

FIGURE 6–42   Shunt resistance and dark current dependence of ambient temperature.

are shown in Figure 6–43. The typical specifications of common diodes are listed in Table 6–4.

When using diodes at low signal levels, noise becomes the limiting factor. Two noise sources are predominant in junction photodiodes: shot noise and thermal noise. Shot noise should be considered at higher signal levels, thermal noise at lower levels. An example of noise calculation and noise effect on signal recognition is given in Chapter 8.

Phototransistors and Photothyristors

Phototransistors and photothyristors are amplifying photodetectors. In contrast to avalanche photodiodes, they have more than one *p-n* junction and the principle of amplification is interaction between junctions, as, for example, in a transistor. The term **photothyristor** includes a variety of photosensitive switching devices, such as photoSCRs (also called LASCRs), photoTRIACs (LATRIACs), and other similar devices. As compared to the detectors described earlier, phototransistors and photothyristors cannot compete with respect to sensitivity, rise time, and linearity. Their advantage is a built-in gain mechanism and, thus, higher responsivity. Despite their shortcomings, they have established themselves as inexpensive and reliable switching and control devices.

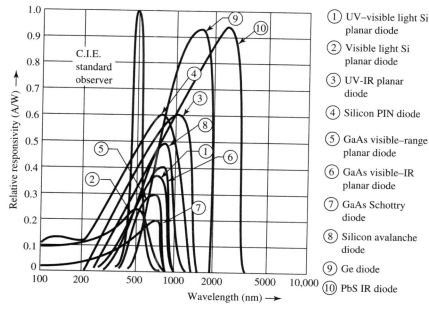

① UV–visible light Si planar diode
② Visible light Si planar diode
③ UV-IR planar diode
④ Silicon PIN diode
⑤ GaAs visible–range planar diode
⑥ GaAs visible–IR planar diode
⑦ GaAs Schottry diode
⑧ Silicon avalanche diode
⑨ Ge diode
⑩ PbS IR diode

FIGURE 6–43   Spectral response of junction diodes.

TABLE 6–4
Specification of Common Photodiodes

Diode Type	Planar Photodiode					Pin Diode	Schottky Diode	Avalanche Diode	Unit
Semiconductor	Si	Ge	GaAsP	InAs	InSb	Si	GaAsP	Si	—
Spectral range	190–1100	700–2000	300–760	500–4500	500–5500	300–1100	190–760	400–1000	nm
Peak response	560–960	1500–1800	640–710	3500	5300	800–960	440–710	800	nm
Relative responsivity	0.3–0.6	1.0	0.3–0.4	$10^{†}$	$200^{†}$	0.5–0.6	0.12–0.72	$40–100^{††}$	A/W
Rise time	0.1–1.0	0.05	0.5–10	0.5	1.0	0.002–0.02	1–30	0.3×10^{-3}–9×10^{-3}	μs
Junction capacitance	20–5000	5–20	2000–6000	10,000	10,000	1–100	0.7–12,000	8–95	pF at 0 V bias
NEP	$0.5–2 \times 10^{-15}$	$0.5–10^{-12}$	$1–5 \times 10^{-15}$	—	—	$8–10^{-11}$	8×10^{-15}	1×10^{-14}	—
D^*	$0.3–3 \times 10^{12}$	$5–8 \times 10^{-10}$	$0.2–1.10^{-12}$	$1–3 \times 10^{9}$	$5–8 \times 10^{-10}$	0.5×10^{15}	0.5×10^{14} 5×10^{13}	1×10^{13}	—
Application	A, B, C, D, E	E	B, I, F	C	—	D, A	F, B	D, A, B	—

A—general radiometry
B—photometry
C—solar cells
D—communication, fiber-optics
E—infrared sensing
F—ultraviolet sensing
†—photovoltaic (V cm²/W)
††—gain

The **phototransistor** is a photodiode followed by a transistor amplifier, as shown in Figure 6-44(a).

Since a transistor already has a built-in *p-n* junction (the base-emitter junction) it can be constructed like a photodiode and the external diode can be eliminated. Therefore, a phototransistor is a transistor whose base-emitter junction is a photojunction. For this reason, the photodiode does not need a base connection and can be built as a two-terminal device, as shown in Figure 6–44(b), or can have a base connection for biasing purposes, as in Figure 6–44(c).

The construction of the phototransistor, as shown in Figure 6–45, is similar to an ordinary transistor except that the base-emitter junction is exposed to radiation, usually through a built-in lens. This design gives a phototransistor directional sensitivity, also shown in Figure 6–45.

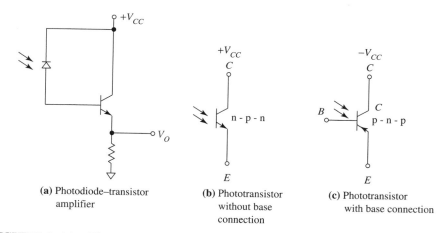

(a) Photodiode–transistor amplifier

(b) Phototransistor without base connection

(c) Phototransistor with base connection

FIGURE 6–44   Phototransistor.

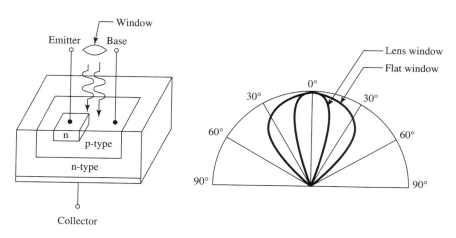

FIGURE 6–45   Phototransistor and polar sensitivity pattern.

The phototransistor behaves like a regular transistor, but the base current is replaced by the photocurrent generated in the photojunction. The steady state relationships in a transistor are

$$i_C = \beta i_B = \alpha i_E \qquad (6\text{--}32)$$

where i_C = collector current (A)
$\quad i_B$ = base current (A)
$\quad i_E$ = emitter current (A)
$\quad \beta$ = current amplification factor
$\quad \alpha = \beta/(\beta + 1)$.

The base current in this case is the photocurrent from the base-emitter photojunction. According to Equation 6–27:

$$i_B = \eta IA\lambda/hc$$

Therefore, the phototransistor collector current is

$$i_C = \beta \eta IA\lambda/hc \qquad (6\text{--}33)$$

The collector current is a linear function of irradiation, as long as the amplification factor β is constant. Beta (β), however, depends on collector current, as shown in Figure 6–46. Therefore the linearity of the phototransistor is limited to a much narrower range than a photodiode or photoconductor.

The phototransistor characteristics, as shown in Figure 6–47, are similar to a regular transistor, only the base current parameter is replaced with irradiance level. The same maximum dissipation and maximum breakdown voltage levels apply.

The responsivity of the phototransistor (RE) in (mA/mW/cm^2) is given for specified black-body radiation, usually at a color temperature of 2870 K, the temperature of a common incandescent bulb. When the detector is used

FIGURE 6–46 Typical variation of current gain with collector current.

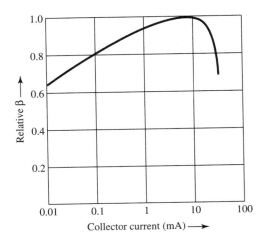

FIGURE 6–47
Phototransistor
characteristics.

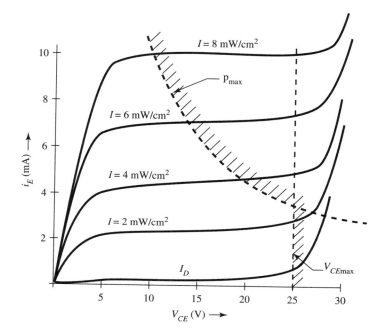

for the visible range of radiation, the responsivity can be converted to mA/lx using the black-body efficacy factor, as given in Figure 1–13, p. 22.

The tolerance of RE is high; -50% to $+100\%$ is quite common. Besides this, RE depends on the irradiance level and ambient temperature. It almost follows the current gain β curve, as shown in Figure 6–46. The temperature coefficient of RE is about $+0.7\%/°C$. Therefore, when designing with a phototransistor, one should consider a wide range of parameter variations.

EXAMPLE 6–10

Design a phototransistor–relay circuit that turns a light bulb OFF when illumination exceeds 400 lx.

Solution
Figure 6–48(a) shows the circuit. The relay with 12 V 100 mA coil is driven by a Motorola MJE3438 transistor with $\beta = 40$ to 160. The phototransistor is Motorola MRD300 with characteristics shown in Figure 6–48(b).

Q_2 base current to activate the relay is

at $\beta = 40$ $i_B = i_C/\beta = 100$ mA/40 $= 2.5$ mA
at $\beta = 160$ $i_B = 100$ mA/160 $= 0.625$ mA

Phototransistor characteristics are given in mW/cm^2 for a 3000 K source (incandescent lamp). The efficacy for a 3000 K black-body radiator is

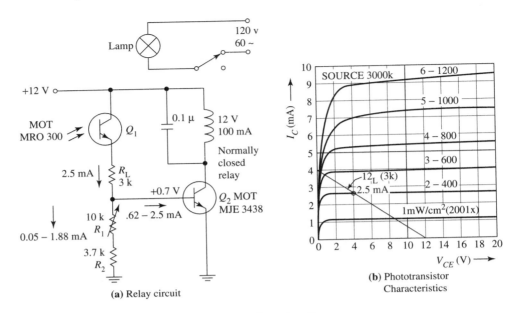

FIGURE 6–48 Phototransistor relay circuit.

K = 20 lm/W (Figure 1–13, pg. 22). Converting mW/cm² to lux we obtain 1 mW/cm² = 20 × 10⁻³/10⁻⁴ lum/m² = 200 lx at 3000 K color temperature.

Thus 2 mW/cm² represents an illumination of 400 lx [Figure 6–48(b)]. Selecting V_{CC} = 12 V and a load line of 3.0 kΩ, the transitor at 400 lx produces photocurrent of 2.5 mA, just sufficient to drive the relay driver with β = 40. With β = 160, only 0.625 mA is required and 2.5 − 0.625 = 1.88 mA has to be by-passed. The base voltage is 0.7 V, thus a shunt resistor R_2 = 0.7/1.88 = 3.72 kΩ has to be included. Since β may have any value between 40 and 160, a series resistor R_1 with a value about 10 kΩ is needed in series with R_2 to adjust the circuit to a tolerance of β.

As in a photodiode, the dark current is the main limiting factor of detection sensitivity. The dark current is a function of ambient temperature and operating condition, as shown in Figure 6–49.

The dark current increases by a factor of 10 for every 20°C increase in ambient temperature. It is also a function of collector emitter voltage.

The rise time of a photodiode is fast. In a phototransistor, rise time is considerably deteriorated due to the base-emitter and base-collector capacitance and the lifetimes of the carriers in the depletion region of the transistor junction. A typical rise time value is a few microseconds. Because of the junction capacitance, the rise and fall times also depend on load resistance. A higher load resistance slows down the rise time and lowers the frequency response of the device. A typical relationship is shown in Figure 6–50.

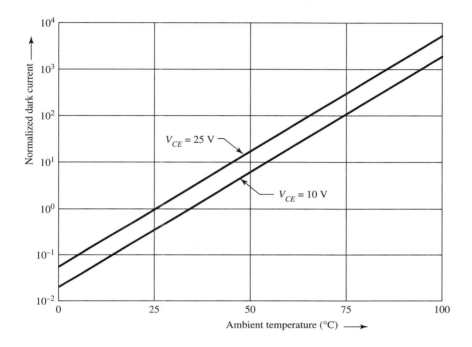

FIGURE 6–49   Phototransistor dark current.

FIGURE 6–50   Typical phototransistor frequency response/load resistance relationship.

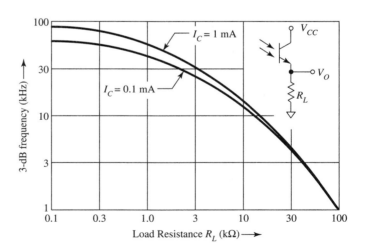

Since the gain from the phototransistor is proportional to the load resistance value, fast response and high gain are contradictory conditions. This problem can be overcome by using a grounded base transistor as a load, as shown in Figure 6–51.

The spectral response of a silicon phototransistor covers the visible-to-low infrared range, as shown in Figure 6–52.

FIGURE 6–51   High frequency phototransistor circuit.

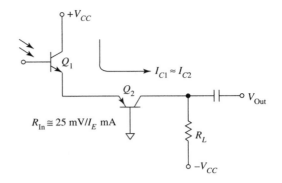

FIGURE 6–52   Typical spectral response of silicone phototransistor.

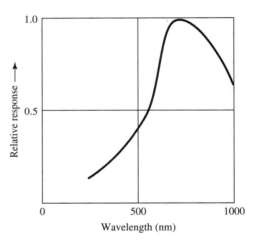

Phototransistors can also be designed in the **photo-darlington** configuration shown in Figure 6–53.

Compared to the phototransistor, the photodarlington has a higher responsivity, achieved at the cost of a much slower response time and decreased linearity.

Besides bipolar phototransistors, **photo field effect transistors** or **photoFETs** are also available. They are constructed with the gate junction exposed to radiation, as depicted in Figure 6–54.

The photoFET combines a photosensitive *p-n* junction with a high impedance, low noise amplifier. The FET is a voltage responding device, thus:

$$\Delta V_o = \Delta V_{\text{GS}} g R_{\text{D}} \tag{6-34}$$

FIGURE 6–53
Photodarlington.

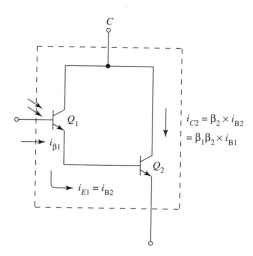

where ΔV_o = output voltage change (V)
ΔV_{GS} = gate-source voltage change (V)
g = FET transconductance (A/V)
R_D = drain resistance (Ω).

ΔV_{GS} is a product of the photocurrent in the gate junction and the gate resistance:

$$\Delta V_{GS} = R_G i_{ph} \tag{6-35}$$

where i_{ph} = gate junction photocurrent (A)
R_G = gate resistance (Ω).

As seen from Equation 6–35 the gate voltage and, therefore, the responsivity of the device can be controlled by the gate resistance. A higher gate

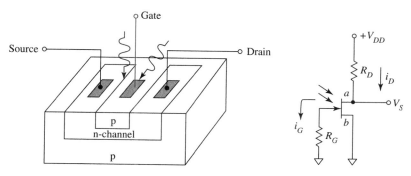

FIGURE 6–54   Photo field effect transistor (photoFET).

resistance will increase the responsivity but at the same time, because of the junction capacitance, will also increase the response time.

Compared to a phototransistor, the photoFET has several advantages. It has higher gain, which is easily adjustable with the gate resistor; its noise performance and frequency response are superior; and it can be temperature compensated using the gate lead. Its disadvantage is that the photoFET has a limited linear range and its response time can be slow when high value gate resistance is used.

Thyristors are electronically controlled switches. With them, one can regulate large amounts of power (in kilowatts) with minimum control power (in milliwatts). In light-controlled thyristors, modest radiation or light power can be used to switch industrial devices. Thyristors are fast-acting devices, and can be used not only for turning a load ON and OFF, but also for controlling the ON/OFF phase of an AC waveform. Therefore, they can act also as proportional control devices.

Two thyristor types are predominant: The **SCR (silicon controlled rectifier)** and **TRIAC**. As the name applies, the SCR is a rectifier, passing current from anode to cathode in only one direction. The rectifier can be turned ON by applying positive current in relation to the cathode to the gate terminal. Once the SCR is turned ON, the gate current is not needed as long as minimum anode-cathode current is maintained. SCR symbols and typical characteristics are shown in Figure 6–55.

The TRIAC is a bidirectional thyristor that can control current flow in both directions, as in AC circuits. It is also a three-terminal device, where the current through terminals T_1 and T_2 can be turned ON by either polarity of the gate signal. As the schematic suggests, the TRIAC is two SCRs connected in parallel in the reverse direction, as shown in Figure 6–56.

The TRIAC, like the SCR, can be turned ON by applying a gate current of either polarity, and it stays ON while the gate current is applied or a minimum current through the device is maintained. Thus, when using a TRIAC

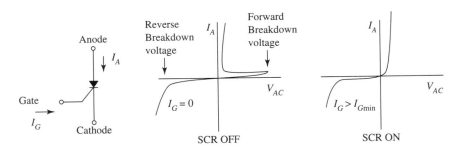

FIGURE 6–55   SCR symbol and characteristics.

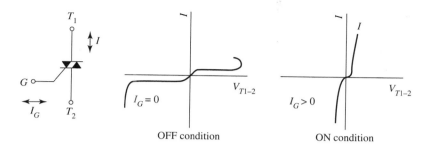

OFF condition   ON condition

FIGURE 6–56   TRIAC symbol and characteristics.

in an AC circuit, the device without gate current turns itself OFF after every zero crossing of the applied voltage. It has to be triggered again during every half cycle of the waveform. The phases of the trigger signal can be used to control the power of the circuit.

Light-actuated SCRs and TRIACs, often called LASCRs (or photoSCRs) and LATRIACs (or photoTRIACs), operate in a similar manner. Instead of the gate current, radiation at the gate actuates the device. These switching devices are very popular in optocouplers, where the actuating source is included in the device. Since the coupling between the source and the SCR or TRIAC is optical, high isolation between input and output circuitry exists and thus low power level, sensitive circuits can be used to actuate high-power devices.

SUMMARY

Chapter 6 discusses radiation detectors. It starts by classifying the detectors by spectral response and operating principles. Further, a set of detector parameters for responsivity, noise equivalent power, detectivity, and D^* is introduced. A short overview of noise, the most important limiting factor in detection systems, is given.

In the section on thermal detectors, thermocouples, bolometers, pyro-electric detectors, and pneumatic- (or Golay-) cells are described and their relative responses and figures of merit given.

Photoeffect detectors are described in more detail. Operating principles and characteristics of vacuum photodiodes, photomultipliers, and microchannels are presented.

Most emphasis is placed on photoconductors and photodiodes. Their operating principles and characteristics are described in detail since they are the most commonly used detectors. Attention is also given to amplifying photodetectors—phototransistors, photodarlingtons, photoFETs, and photothyristors.

RECOMMENDED READING

Dereniak, E. L., and Crowe, D. G. 1984. *Optical Radiation Detectors.* New York: John Wiley and Sons.

Wilson, J., and Hawkes, J. F. B. 1969. *Optoelectronics: An Introduction.* New York: Prentice Hall.

Chapel, A. 1978. *Optoelectronics: Theory and Practice.* New York: McGraw-Hill Book Company.

Hewlett-Packard. 1981. *Optoelectronics Fiber-Optics Application Manual.* New York: McGraw Hill Book Company.

Driscoll, W. G., Editor. 1978. *Handbook of Optics.* New York: McGraw-Hill Book Company.

PROBLEMS

1. Using Figure 6–13, calculate the gain of a six-stage photomultiplier with NAE-type dynodes and 200 V dynode voltage. Calculate the dark current when a metal cathode (as on Example 6–2) is used ($i_T = 11.0$ fA).

2. Calculate λ_{max} wavelength of doped germanium with an energy gap of 0.06 V.

3. A photoconductor with $\alpha = 0.75$ has resistance of 120 kΩ at illumination of 100 lx. Find the resistance at 50 and 500 lx.

4. In an LED intensity control circuit [Figure 6–57(a)], a photoconductor with the resistance value of $R_C = 2.5$ kΩ at 10 lx illumination and $\alpha = 0.6$ sends 2 mA through the LED. Find the value of R_1 and LED current at 1000 lx illumination.

5. A solar cell has $V_{L0} = 0.6$ V and $i_{L0} = 200$ mA. At optimal loading it delivers 160 mA into the load. Find optimal load resistance, cell output voltage, and output power, and design an array to charge a storage battery with 6.6 V and 0.5 A.

6. Design a four-range illumination meter with 300, 100, 30, and 10 fc full scale. Use a photodiode with short circuit photocurrent of 0.5 μA at 100 lx illumination. Use the circuit shown in Figure 6–34(b) and meter with full scale indication at 1.00 V.

7. Design the same illumination meter as in Problem 6, using the circuit shown in Figure 6–34(a). Select $R_L = 10$ kΩ.

8. Design a photographic incident light meter with logarithmic scale. The full indication shall be at 10,000 fc. Use photodiode with the response of 0.5 μA at 100 lx, the dark current of 10 pA, and a meter with full-scale indication of 1.00 V. Calculate the indication at typical TV stage illumination of 1,500 fc.

9. Select the right junction diode for the following applications:
   a. to detect infrared radiation in the 2–3 μm range
   b. fast response diode for fiber-optics application
   c. diode with high responsivity
   d. diode to detect low-level signal.
10. Design a phototransistor punch-card reader that activates a TTL flip-flop. Radiation luminance is 1 mW/cm^2 and V_{CC} = 12 V. Use the transistor with characteristics as shown in Figure 6–48(b).
11. Design a punch-card reader that activates a TTL flip-flop using a photo-FET. PhotoFET characteristics are responsivity 0.3 A/V, transconductance 3 mA/V, and sensitive area 0.4 cm^2.
12. Calculate the required luminance to produce a 5.0 V drop in a 1.0 kΩ load resistor in photodarlington with $\beta_1\beta_2$ = 5,000, responsivity of 0.3 A/V, and junction area of 10 mm^2.

7
Optical Sensors and Optocouplers

Most optoelectronics applications involve manipulating a source and a detector. By measuring or controlling the flow of flux from the source to the detector, a variety of tasks can be accomplished, from operating a simple punch-card reader to managing complex communication, imaging, and measurement systems. In fact, the applications are so numerous that only the basic optoelectronics sensing and measuring techniques are covered in this chapter. Special attention is also given to a relatively new group of components, the optocouplers.

7–1 PRINCIPLES OF OPTICAL SENSING

An optocoupled link is a source and a receiver that are coupled by radiant flux. An efficient link produces maximum receiver response from a source with minimum power, requiring that the maximum amount of source flux be concentrated at the receiver. Also, the source and receiver wavelength characteristics must be matched. These coupling characteristics can be measured with three coefficients: axial, angular, and wavelength alignment efficiencies.

Coupling Efficiency

In Chapter 1 the terms **optical transfer function** and **numerical aperture** were introduced. They describe the efficiency of optical coupling in terms of total flux transfer from the source to the detector. **Axial alignment**

efficiency is a similar term, limited to source, detector, and aperture geometry. It is described with the help of Figure 7–1.

The aim of the designer is to concentrate all the flux through the aperture into the detector. In the case where the flux cone cross-section area A_S equals the detector area A_D, the axial alignment efficiency equals one. Assuming circular aperture and detector area, we can define:

$$\eta_A = A_D/A_S = D_D{}^2/D_S{}^2 \qquad (7\text{–}1)$$

where η_A = axial alignment efficiency
   A_S = flux cone area at the detector plane (mm^2)
   A_D = detector area (mm^2)
   D_S = flux cone diameter at the detector plane (mm)
   D_D = detector diameter (mm).

Three conditions, shown in Figure 7–2, are possible. In 7–2(a) the detector area is smaller than the flux cone area. In this case, the alignment efficiency is less than one and can be calculated using Equation 7–1. In Figure 7–1(b) the detector area is greater than the flux cone area, and therefore all flux is received by the detector. The alignment efficiency in this case is always one. Moreover, this condition tolerates some misalignment of the detector with respect to the optical axes, an important practical requirement. However, there is a limit to how concentrated the flux on the detector surface can be. Some detectors exhibit nonlinearities under these conditions. Case (c) demonstrates misalignment of an otherwise perfect cone and detector area relationship. It demonstrates a practical case of good design in consideration of production tolerances. The efficiency in this case is the ratio of intersecting

FIGURE 7–1   Source-detector geometry.

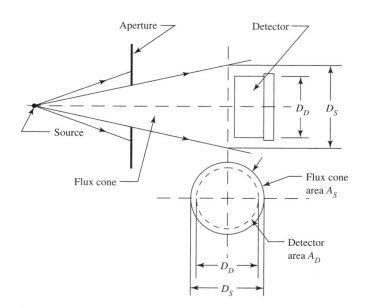

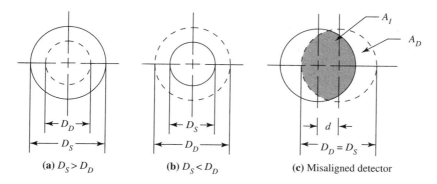

(a) $D_S > D_D$ **(b)** $D_S < D_D$ **(c)** Misaligned detector

FIGURE 7–2   Possible source-detector alignment conditions.

area A_I to detector area A_D. Depending on the misalignment, displacement d, the efficiency can be read from the graph in Figure 7–3.

EXAMPLE 7–1

A 2-mW flux from a source with a cone diameter of 2.5 mm is directed to a receiver with a 2.5 mm diameter. In the setup a 0.5 mm alignment displacement is made. How much flux is received by the detector?

Solution
The alignment error is $0.5/2.5 = 0.2$. From the graph in Figure 7–3, the alignment efficiency is 0.74. Thus, $0.74 \times 2.0 = 1.48$ mW flux is coupled into the detector.

FIGURE 7–3   Alignment displacement effect of alignment efficiency.

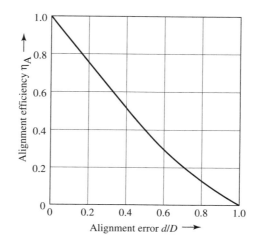

The detector surface should be perpendicular to the optical axes. Due to production tolerances or other reasons this may not be the case, thus reducing the flux coupling. This effect can be expressed in **angular misalignment efficiency,** depicted in Figure 7–4.

The angular misalignment efficiency can be expressed as

$$\eta_\alpha = \cos \alpha \tag{7–2}$$

where η_α = angular misalignment efficiency
α = deviation of detector surface from the perpendicular to the optical axes (deg).

This effect is usually negligible since an angular alignment error of 10° causes a flux loss of only 1.5%.

The **wavelength alignment efficiency** can, on the other hand, be a considerable factor, especially when semiconductor detectors and sources are used. The spectral response of the detector may not match the spectral output of the source. A typical condition is shown in Figure 7–5.

As seen from the graph, the spectral response of the source and receiver may not be the same. Thus the detector responds to only a fraction of the source flux. In Figure 7–5, curve $f(S)$ represents the relative source spectrum and curve $f(D)$, the detector response. Multiplying the values of curve $f(S)$ by $f(D)$, the curve $f(S)f(D)$ can be constructed. This represents, at a given wavelength, the fraction of the source flux detected by the detector.

The area under product curve A_P represents the total relative power detected by the detector. The area under the source curve represents the total relative power radiated by the source. The wavelength alignment efficiency is the ratio of these areas.

$$\eta_\lambda = A_\mathrm{P}/A_\mathrm{S} = \int_\lambda f(S)\,f(P)\,d\lambda \Big/ \int_\lambda f(S)\,d\lambda \tag{7–3}$$

where η_λ = wavelength alignment efficiency
$A(P)$ = total relative power received by the detector
$A(S)$ = total relative power radiated by the source.

Numerical integration, similar to calculating efficacy, as described in Chapter 1, must be carried out since mathematical functions for source and

FIGURE 7–4   Angular misalignment.

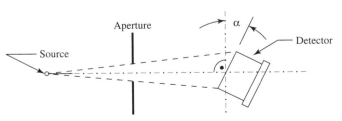

FIGURE 7–5   Source and detector wavelength response matching.

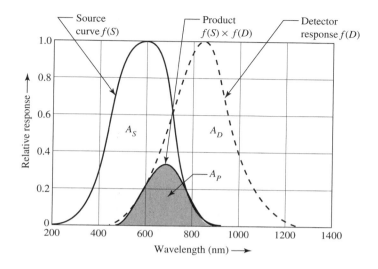

detector response do not exist. Relative spectral responses for most sources are given in earlier chapters in this book and also in component data sheets. In Table 7–1, a few examples of wavelength alignment efficiency are given for the human eye (peak response at 550 nm) and a silicon phototransistor (peak response at 800 nm).

As seen from Table 7–1, a silicon phototransistor, having a much wider spectral response with the peak at a higher wavelength, is a much more efficient detector for given sources than the human eye.

Wavelength alignment efficiency does not apply with wide spectral response detectors, such as thermal detectors, since their response usually covers the entire source spectrum. However, when the optical path goes through a medium that has a wavelength dependent transparency, its effect on the receiver response is calculated in a similar manner.

TABLE 7–1
Wavelength Alignment Efficiency for Human Eye and Silicon Phototransistor for Several Sources

Source	Human Eye	Silicone Phototransistor
Tungsten Lamp 2000 K	0.003	0.16
Tungsten Lamp 2800 K	0.30	0.27
Neon Lamp (600 nm)	0.35	0.70
GaAs IRED LED (940 nm)	0	0.80
GaAlAs IRED LED (880 nm)	0	0.98
Sun	0.16	0.50

The **total coupling efficiency** can be expressed by the following equation:

$$\eta_T = \eta_A \eta_\alpha \eta_\lambda \qquad \qquad \textbf{(7–4)}$$

where η_T = total coupling efficiency.

EXAMPLE 7–2

Design a TV remote control unit and find the angle range of its operation at a 10 ft distance.

Solution

A TV remote control unit has an infrared transmitter that sends a digital pulse train to the receiver at the TV set. Let us select a 940 nm 100 mW infrared LED with polar intensity profile $I_\theta = I_0 \cos^3\theta$. Receiver: silicone photodiode with $RE = 0.5$ mA/mW/cm² for 2800 K tungsten source with dark current 50 μA. Efficiency for 2800 K is 0.30 and for 920 nm radiation, 0.80. The configuration is shown in Figure 7–6.

From the equations in Table 1–3, the maximum intensity of the LED is

$$I_0 = \phi(n + 1)/2\pi = (100 \times 4)/(2 \times 3.14) = 63.7 \text{ mW/sr}$$

At 10 ft = 3.05 m and transmitter facing the receiver ($\alpha = 0$), the incidence at the receiver is

$$E_0 = I_0/d^2 = 63.7/3.05^2 = 6.85 \text{ mW/m}^2 = 6.85 \times 10^{-4} \text{ mW/cm}^2$$

Receiver responsivity is given for a 2800 K tungsten source. Knowing the efficiency at 2800 K and 920 nm, we can convert

$$RE_{920nm} = RE_{2800K} \times (\eta_{920}/\eta_{2800K}) = 0.5(0.8/0.3) = 1.333 \text{ mA/mW/cm}^2$$

Thus, the receiver output is

$$i_0 = E \times RE_{920nm} = 1.333 \times 6.85 \times 10^{-4} = 9.13 \text{ } \mu\text{A}$$

The signal to dark current ratio is

$$i_0/i_D = 9.13/0.050 = 18.3$$

in a reasonable range.

Assuming that a minimum ratio of 10 is required for reliable operation, we can calculate the angular efficiency

$$\eta_\alpha = 10/18.3 = 0.546$$

This corresponds to the maximum angle from the perpendicular

$$\alpha_{max} = \arccos 0.546 = 56.9°$$

Applications for optical sensors are numerous: card or tape readers, shaft position encoders, tachometers, bar-code scanners, smoke and burglar

FIGURE 7–6   Example 7–2
configuration.

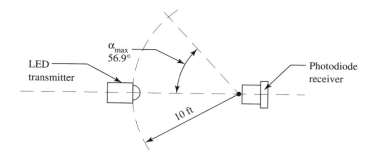

alarms, and edge sensors are some examples. Optical sensors can be grouped into two types: reflective and transmissive systems. As shown in Figure 7–7, both systems have a source and a receiver. A beam of radiation is directed from the source to the detector. The beam is modulated by the medium or object between the source and the receiver. In the case of a reflective system, the beam is reflected from the object surface. In a transmissive system, the beam passes through, or is interrupted by, the object. In either case, the beam can be interrupted, digitally coded, or amplitude modulated, as shown in Figure 7–7.

The task of the designer is to select the components to accomplish the design's purpose. A few basic design principles are common to most optical sensor systems.

Edge and Bar Detection

Edge and bar detection techniques are common to all devices that depend on interruption of the beam. Object counters, speedometers, and bar-code scanners are typical applications of this characteristic. In edge detection, a clearly

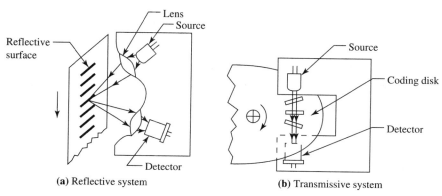

(a) Reflective system   (b) Transmissive system

FIGURE 7–7   Optical sensor systems.

defined opaque object in a transmissive system, or a black nonreflective surface in a reflective system, interrupts the optical path, decreasing the receiver output from a maximum to a minimum, as shown in Figure 7–8.

The transfer of detector output takes place at the interval where the edge moves across the cross section of the beam. Therefore, the precision on the edge location depends on the beam diameter. When the detector output change is measured from 10 to 90% maximum output, the change takes place at a distance of 69% of the beam diameter. For many applications, such as simple object counting and burglar alarms, any beam diameter will work. However, applications such as bar-chart reading and position and angle decoders require a high level of precision. A typical application is counting parallel lines or bars, as shown in Figure 7–9.

As demonstrated, the detector output depends on the beam cross section and bar-spacing relationship. When the cross section is less than the spacing between bars (case a), the detector output varies from maximum to minimum, or in other words, the output is 100% modulated. In cases (b) and (c), the beam diameter is increasingly wider compared to the bar spacing, and the detector output swing is less (the output modulation is decreased). We can define the modulation of the output signal as follows:

$$\text{MOD} = (m_{\max} - m_{\min})/(m_{\max} + m_{\min}) \qquad \textbf{(7–5)}$$

where MOD = ratio of output signal modulation
$m_{\max}$ = maximum amplitude of the output signal (V, A)
$m_{\min}$ = minimum amplitude of the output signal (V, A).

Modulation is a measure of the quality of pattern recognition. High output signal modulation means that the pattern is well recognized, and at low modulation the recognition of the pattern may be disturbed by noise and

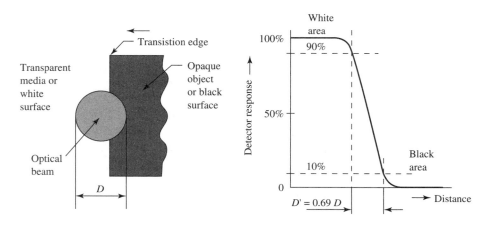

FIGURE 7–8   Edge detection with optical sensor.

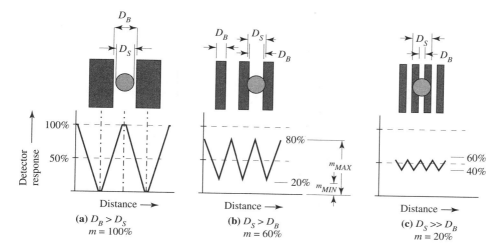

FIGURE 7–9   Detector output modulation for different patterns.

other side effects. By combining the output modulation with the pattern characteristics (or **pattern spatial frequency**), a measure to evaluate pattern recognition, the **modulation transfer function (MTF),** can be defined. The spatial frequency is measured in "line pairs per millimeter":

$$F = 1 \text{ mm}/2D_B \qquad (7\text{–}6)$$

where F = spatial frequency (line pairs per mm)
D_B = width of a bar in a pattern where bar spacing equals the bar width (mm).

The modulation transfer function, or MTF, is specified as the percent of output modulation at a given spatial frequency. Usually it is given in graph form, as seen in Figure 7–10, the MTF graph for the HP HEDS-1000 High Resolution Optical Reflective Sensor.

Besides the spacial frequency of the pattern, the MTF also depends on the absorbance of the black bars, reflectivity of the spaces between the bars, and distance of the detector from the pattern.

EXAMPLE 7–3

A typical bar code test pattern has a bar width of 0.19 mm (0.0075 in.). What is the MTF using an HP HEDS-1000 sensor?

Solution

A line width of 0.19 mm corresponds to a spatial frequency of $1/2 \times 0.19 = 2.63$ line pairs per millimeter. As seen in the graph in Figure 7–9, it produces an MTF of 57%. Using proper digital circuitry, this is satisfactory for recognizing a code.

FIGURE 7–10   Typical
modulation transfer function.

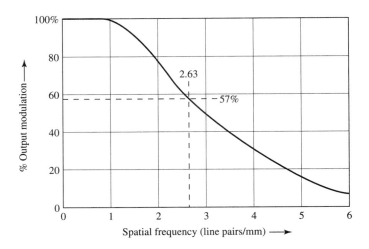

There are two ways to create optical sensing systems with the highest possible resolution: using auxiliary patterns or reducing the cross section area of the sensing beam.

Auxiliary patterns can be employed to detect tight, uniformly spaced bar patterns that otherwise would require a very small diameter test beam. Measuring linear movement or determining shaft rotation are common applications of this technique. A typical example is shown in Figure 7–11.

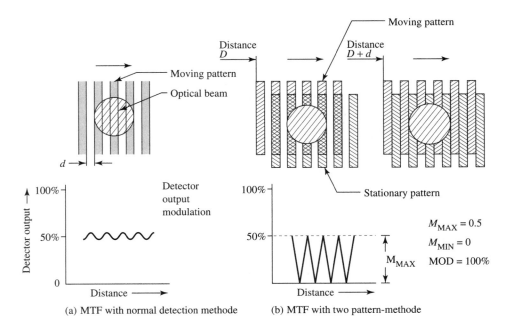

FIGURE 7–11   Sensor with stationary and moving pattern.

Case (a) depicts the output modulation graph of a tightly spaced pattern detected with a wide diameter beam. It produces a low and unusable detector output modulation. To improve the modulation, a stationary bar pattern of the same spacing can be placed between the moving pattern and the detector. When the patterns are aligned, about 50% of the flux passes to the detector. If the moving pattern moves a bar width, the flux is completely blocked. The resulting detector output is 100% modulated but has only half of the maximum amplitude, as shown in case (b). This method increases system resolution, but reduces the output amplitude by one-half.

Besides counting the bars, it is often important to determine the direction of the pattern movement. A typical application would be fixing the location of the work table of a machine. In this case, the count should be increasing when the table moves in one direction and decreasing when it moves in the opposite direction. This can be accomplished by using two sensors, as shown in Figure 7–12. The phase relationship between the outputs of the detectors changes, depending on the direction of movement of the pattern. Figure 7–12 shows a simple FF circuit that diverts the count pulses to UP or DOWN inputs of an UP/DOWN counter, depending on the phase condition of sensor A and B outputs.

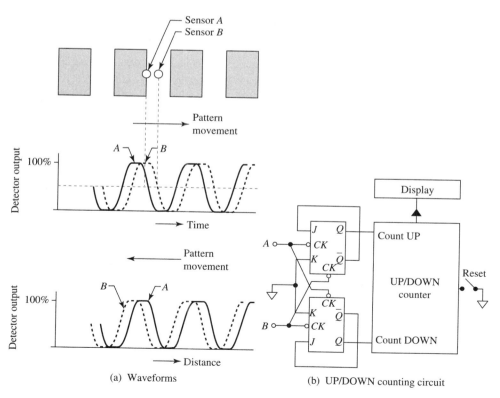

(a) Waveforms

(b) UP/DOWN counting circuit

FIGURE 7–12   Sensor arrangement and counter for UP/DOWN counting.

EXAMPLE 7–4

Designing a sensing system for a digital readout bathroom scale.

Solution

A bathroom scale has a 12-inch diameter disk whose rotation is proportional to the weight on the scale. The task is to design a digital UP/DOWN readout system to indicate disk. Let us assume that the disk rotated 360 degrees when 280 lbs weight is on the scale. Thus, we will punch 280 slots at the periphery of the disk, as shown in Figure 7–13.

At 5.5 inch radius the slot period is

$$d = \pi R/280 = 0.0617 \text{ in.} = 1.567 \text{ mm}$$

For UP/DOWN counting we need two sensors a half period apart, as shown in Figure 7–13(b). For high MTF, either the source flux cone or the detector area has to be less than the slot width. Detectors have a wider response area than the slot width of 0.78 mm. LEDs with very narrow beam pattern are available, such as HP HMT-6000, which is a 700 nm emitting LED with half angle of 8 degrees. The spot diameter (D_S) should not be greater than 0.75 mm. This allows us to calculate the LED distance (d) from the slot:

$$D_S/2d = \tan \theta_{1/2} \quad \text{or} \quad d = D_S/2 \tan \theta_{1/2} = 0.75/2 \tan 8° = 2.67 \text{ mm}$$

The distance to the photodiode is not critical since the entire flux is modulated by the slot.

When the bar pattern is very tight, the location of sensors in a correct phase relationship becomes difficult. The use of moiré patterns can solve this problem.

A **moiré pattern** is created when the spacing of the stationary bar pattern is different from the spacing of the moving bar pattern, as shown in Figure 7–14.

Over a distance where the stationary and moving bar count differs by one, a **moiré period** is formed. Locating sensors with respect to this distance

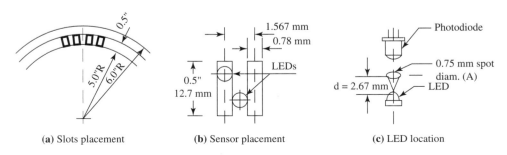

(a) Slots placement   (b) Sensor placement   (c) LED location

FIGURE 7–13   Digital readout sensing system.

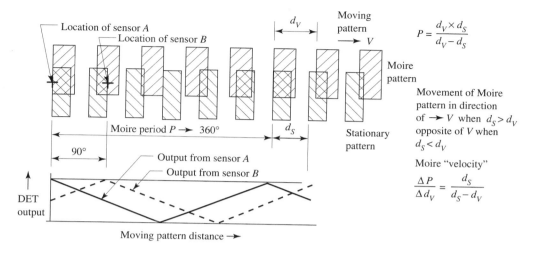

FIGURE 7–14   Directional information from moiré pattern.

allows any phase relationship between the two sensors, providing more tolerance in sensor location, compared to one bar period. Moiré patterns will move in the direction of the moving pattern when the bar spacing on the stationary pattern is greater than the moving pattern spacing. Opposite movement is achieved under the reverse condition.

Sensing Methods Using Lenses

As described, an auxiliary pattern can be used to increase resolution when a wide detector beam is used. The resolution can also be increased by decreasing the source beam cross section. This design requires the use of lenses, and is thus a more costly and complicated solution. A typical application, a precision edge detector using confocal coupling, is shown in Figure 7–15.

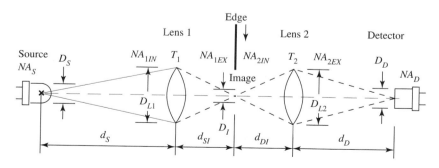

FIGURE 7–15   Confocal edge sensing system.

The designer must select lenses and determine their location to form a sufficiently small image of the source at the edge crossing plane and then calculate the optical transfer function (OTF) for the system. The equations for the calculations are given in the Chapters 1 and 2. Following is a demonstration of their application.

Lens 1 forms an image of the source at the edge detection plane. The diameter of the image determines the precision of the edge detection. The OTF of the system is determined by the flux transferred from the source to the detector through lenses 1 and 2.

Using the quantities as defined in Figure 7–15, the image diameter at the edge plane can be written as:

$$D_I = MD_S \tag{7–7}$$

where M = magnification/reduction factor of lens 1.

$$M = d_{SI}/d_S \tag{7–8}$$

Knowing the desired M, the focal length of the lens and the actual distances d_{SI} and d_S can be obtained from Equations 2–12 to 2–19, pp. 64 and 65.

Using the term **numerical aperture (NA),** we can write the OTF of the system, assuming that all flux from lens 2 is received by the detector:

$$OTF = \phi_D/\phi_S = (NA_{1IN}/NA_S)(NA_{2IN}/NA_{1EX})T_1 T_2 \tag{7–9}$$

where ϕ_S = total source flux (W, lm)
ϕ_D = flux received by the detector (W, lm)
T_1 = transmission coefficient of lens 1
T_2 = transmission coefficient of lens 2.

The first term of the equation describes the flux transfer from the source to lens 1, the second term from lens 1 to lens 2. When two similar lenses are used and they are located so that $d_{SI} = d_{DI}$ and $NA_{1EX} = NA_{1IN}$, the equation simplifies to

$$OTF = (NA_{1IN}/NA_S)T_1 T_2 \tag{7–10}$$

This expression says that all flux, with the exemption of the transmission loss in the lenses, received by lens 1 is transferred to the detector. If the source has a Lambertian intensity distribution, the OTF equation simplifies further:

$$OTF = NA_{1IN}^2 T_1 T_2 \tag{7–11}$$

To find the flux transfer from the source to lens 1, the equations given in Table 1–3, pp. 33, can also be used.

EXAMPLE 7–5

Design an edge detector with 0.5 mm resolution.

Solution

For 0.5 mm resolution the flux cone diameter at the edge plane should be 0.5 mm or less. This can be done using a simple lens system shown in Figure 7–16.

The source, HP HEM-800 LED, emits 700 nm, $I_0 = 250$ μW/sr, $\theta_{1/2} = 8°$, with radiation area 1 mm diameter. The receiver is a Hamamatsu SR26-44BQ silicone diode, $RE = 0.35$ A/W at 720 nm, and sensitive area 3.6 × 3.6 mm.

For cone diameter 0.5 mm the reduction ratio is $M = 0.5$ and $i/o = 0.5$. Selecting $o = 20$ mm, we can calculate lens focal length using Equation 2–18.

$$f = (o + i)/(2 + M = 1/M) = (20 + 10)/(2 + 0.5 + 1/0.5) = 6.67 \text{ mm}$$

Selecting lens diameter $D = 10$ mm, we can calculate the flux cationed by the lens. The flux cone angle is

$$\theta = \arctan(5/20) = 14.0 \text{ deg}$$

The LED radiation pattern is $I_\theta = I_0 \cos^n \theta$. From the half angle using Equation 1–19, the n is:

$$n = |0.31/\text{logcos } \theta 1/2| = 71$$

From the equations in Table 1–3 the flux is

$$\phi = 2\pi I_0 (1 - \cos^{n+1}\theta/(n + 1)) = (2\pi \times 250)(1 - \cos^{72} 14.0°)/72 = 19.3 \text{ } \mu\text{W}$$

Assuming the lens transmission coefficient $T = 0.9$, the flux into receiver is

$$\phi_R = \phi T = 19.3 \times 0.9 = 17.4 \text{ } \mu\text{W}$$

Since the spectral response of the diode matches the LED output, $\eta_\lambda = 1.0$ and

$$i_D = \phi RE = 17.4 \times 0.35 = 6.09 \text{ } \mu\text{A}$$

To capture the entire flux, the diode must be placed where the flux cone diameter is 3.0 mm. Thus the diode distance from the edge plane is:

$$d = i(D_D/D) = 10 \times (3.0/10) = 3.0 \text{ mm}$$

FIGURE 7–16   Edge sensing system.

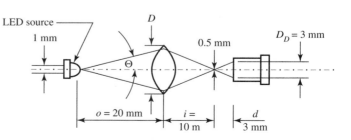

The system described in Figure 7–16 is a transmissive system. The same principle applies also to a reflective system, as shown in Figure 7–17.

In a reflective system, the lens S projects the source image onto the target surface. When the target has a specular surface (that is, the target is a mirror), the image is reflected through lens 2 to the detector. The system can be designed using the OTF just described.

There are, however, two more parameters to consider: the reflectivity loss at the target surface and the alignment efficiency of the projected image. When the target has a diffusing surface, the OTF calculation can also be carried out, as long as the diffusion characteristics of the target surface are known. The system can also be analyzed in terms of the illuminance or incidance. The latter method is often preferred because the detector responsivity is usually given with the incidance as the parameter. A typical analysis based on incidance or illumination calculation is shown below.

For incidance calculation often a parameter, **F-Number, (F-NUM),** is used. This parameter is popular in photography, a field that deals with forming an image. As discussed in Chapter 2, the $F\text{-}NUM$ is defined as

$$F\text{-}NUM = f/D = 1/2NA_f \tag{7–12}$$

where f = focal length of a lens (mm)
$\quad D$ = diameter of lens opening (mm)
$\quad NA_f$ = numerical aperture when lens is one focal length away from the source.

The $F\text{-}NUM,$ like the $NA,$ describes the lens' ability to transfer the flux. Their relationship, however, is inverse: a greater NA means that the lens can pass more flux, while a greater $F\text{-}NUM$ means the lens passes less flux.

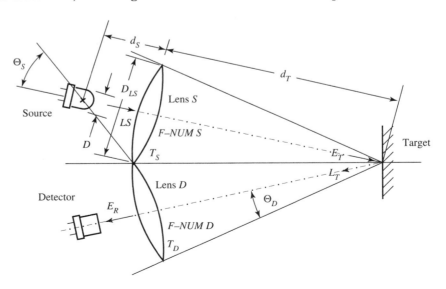

FIGURE 7–17   Reflective optical sensor.

Without a lens and using a point source, the target incidence or illumination can be calculated:

$$E_T = I_S/d_T^2 \tag{7–13}$$

where E_T = target incidence or illumination (W/m² or lx)
$\quad I_S$ = source intensity (W/sr or cd)
$\quad d_T$ = source-target or source-lens distance (m).

When the source is a circular area, the relationship is

$$E_T = L_S \pi D^2/4d_T^2 = L_S \omega_S \tag{7–14}$$

where D = source diameter (m)
$\quad L_S$ = source sterance or luminance (W/sr/m² or cd/m²)
$\quad \omega_S$ = angular range of the source subtended at the target (sr).

When a lens is used between the source and target, as shown in Figure 7–17:

$$E_T = \pi L_S T/4(F\text{-}NUM)^2(1 + d_T/d_S)^2 = \pi L_S T/4(F\text{-}NUM)^2(1 + M)^2 \tag{7–15}$$

where M = magnification factor $M = d_T/d_S$
$\quad T$ = transmission coefficient of the lens.

When the source is a long distance away so that $d_T \approx f$ and $d_S \gg f$, Equation 7–15 simplifies to

$$E_T = L_S T \pi D_{LS}^2/4d_T^2 = L_S T A_{LS}/d_T^2 = L_S T \omega_{LS} \tag{7–16}$$

where D_{LS} = diameter of the source lens (m)
$\quad A_{LS}$ = area of the source lens (m²)
$\quad \omega_{LS}$ = angular range of the source lens subtended at the target (sr).

This equation is similar to the point source relationship where the source has an intensity of $L_{Sm} T A_{SD}$. Since $\omega_{LS} > \omega_S$, introducing a lens between the target and source increases the target illuminance.

EXAMPLE 7–6

Calculate target illumination for a target that is 1 m from a source that has the following characteristics:

    a. a point source with intensity of 50 mcd
    b. a circular area source with the diameter of 5 mm and intensity of 50 mcd over the area
    c. a source as in case (b), but using a 20 mm diameter lens to focus the flux to the target

The example is illustrated in Figure 7–18.

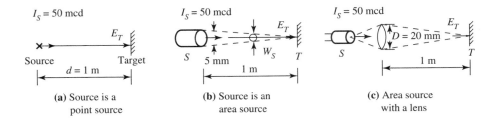

FIGURE 7–18   Configurations for Example 7–5.

Solution (a)
From Equation 7–13:

$$E_T = I_S/d^2 = 50 \times 10^{-3}/1^2 = 50 \text{ mlx}$$

Solution (b)
Source area is

$$A_S = \pi D_S^2/4 = \pi \times 5^2/4 = 19.6 \text{ mm}^2 = 19.6 \times 10^{-6} \text{ m}^2$$

Source luminance is thus

$$L_S = I_S/A_S = 50 \times 10^{-3}/19.9 \times 10^{-6} = 2.55 \times 10^3 \text{ cd/m}^2$$

The angular range of the source is

$$\omega_S = A_S/d^2 = 19.6 \times 10^{-6}/1^2 = 19.6 \times 10^{-6} \text{ sr}$$

Using Equation 7–14 we obtain

$$E_T = L_S\omega_S = (2.55 \times 10^3)(19.6 \times 10^{-6}) = 50.0 \times 10^{-3} = 50 \text{ mlx}$$

Since the intensity of the area source is the same as the intensity of the point source, the illumination is the same.

Solution (c)
The lens area is

$$A_L = \pi D_L^2/4 = \pi 20^2/4 = 314 \text{ mm}^2 = 314 \times 10^{-6} \text{ m}^2$$

The angular range from the target to the lens is

$$\omega_L = A_L/d^2 = 314 \times 10^{-6}/1^2 = 314 \times 10^{-6} \text{ sr}$$

Thus, from Equation 7–16, assuming $T = 0.95$, we have

$$E_T = L_S T\omega_L = (2.55 \times 10^3)(0.95)(314 \times 10^{-6}) = 760 \text{ mcd}$$

Introducing the lens improves the target illumination by a factor of 16.

___

At the target, a portion of the incident flux is reflected back to the detector. How much, depends on the type of reflector used. If the target has a specular reflecting surface, as in a burglar alarm, the target image acts as a source with the sterance of

$$L_{Tspec} = RE_T/\omega_R \qquad (7\text{–}17)$$

where L_{Tspec} = sterance of a target with specular surface (W/st/m^2 or lm/sr/m^2)
R = reflection coefficient of the surface
E_T = target irradiance (w/m^2 or lm/m^2)
ω_R = angular range of exit flux (sr).

In the case of a specular target area, the entrance and exit flux angular ranges are the same and the image is projected to the detector.

When the target surface is diffusing (that is, the target is a Lambertian reflector, as in the case of a bar-code reader), the angular range of the reflected flux equals π and

$$L_{Tdif} = RE_T/\pi \qquad (7\text{–}18)$$

where L_{tdif} = sterance of a target with diffusing surface with Lambertian profile (W/sr/m^2 or lm/sr/m^2).

The effective reflection coefficient of a diffusing reflector R_{dif} is therefore

$$R_{dif} = R/\pi \qquad (7\text{–}19)$$

As a rule, specular reflectors are much more efficient, with a reflection coefficient above 0.9. A diffusing target may have a reflection coefficient below 0.8 which, with the additional coefficient of π, reduces the total coefficient to only 0.25.

The radiant incidence to the detector in a reflective system, as shown in Figure 7–17, can be calculated using $F\text{-}NUM$ to describe the lenses:

$$E_R = L_S\pi T_S T_D R/(F\text{-}NUM_S)^2 (F\text{-}NUM_D)^2 (1 + d_T/d_S)^2 (1 + d_D/d_T)^2 \qquad (7\text{–}20)$$

When the target is far away, the equation simplifies to

$$E_R = (L_S R/\pi)(T_S A_{LS}/d_T{}^2)(T_D A_{LD}/d_D{}^2) \qquad (7\text{–}21)$$

The terms used above are explained in Figure 7–14 and in earlier equations.

The problem can also be solved using the numerical aperture and OTF. In this case, the OTF from the source to the detector is

$$OTF = NA_S{}^2 T_S NA_D{}^2 T_D R_T \qquad (7\text{–}22)$$

EXAMPLE 7–7

Design a reflective counter that counts the customers passing through a store entrance. The configuration of the system is shown in Figure 7–19. Calculate the incidence at the receiver surface and the receiver output current for a system with a specular or diffusing reflector and with or without a receiver lens.

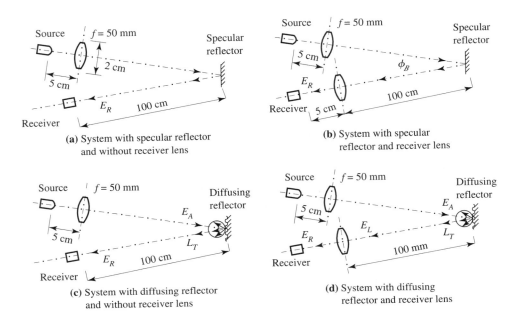

FIGURE 7-19   Configurations for Example 7-7.

Solution

Select

Source: TI type TIL33 GaAs infrared emitting diode.

$$\phi = 5 \text{ mW} \quad \text{at} \quad 940 \text{ nm}; \; \theta_{1/2} = 40°;$$

Receiver: TI type 1N5724 NPN planar silicon phototransistor.

$$RE = 0.3 \text{ mA/mW/cm}^2 \text{ for tungsten light 2800 K};$$

$$\text{Dia} = 1.52 \text{ mm}; \; \eta_{2800K} = 0.30; \; \eta_{940nm} = 0.8;$$

Lens: $f = 50$ mm; Dia $= 20$ mm.

Coupling from the source to the source lens: The diode polar intensity profile is $I_\theta = I_0 \cos^n \theta$; the exponent n can be calculated from Equation 1–20:

$$n = |0.301/\text{logcos} \, \theta^{1/2}| = |0.31/\text{logcos} \, 40°| = 2.68.$$

Thus, the source intensity profile and maximum intensity are

$$I_\theta = I_o \cos^{2.68}\theta \quad \text{and} \quad I_o = \phi(n + 1)/2\pi = 5.0(2.68 + 1)/2\pi = 2.93 \text{ mcd}$$

The coupling angle from the source to source lens is:

$$\theta_S = \arctan(20/2 \times 50) = 11.31°$$

and, using the equations in Table 1–3, the flux coupled into the lens is

$$\phi_L = 2\pi I_o(1 - \cos^{n+1}\theta_S)/(1 + n) = 2\pi \times 2.93 \times (1 - \cos^{3.68}11.31°)/3.68$$

$$= 0.35 \text{ mW}$$

Since the source is at the focal point of the lens, the lens forms a collimated beam. The reflected beam flux, considering the lens transmission coefficient $T = 0.95$ and the reflection coefficient of the mirror $R = 0.90$, is

$$\phi_B = \phi_L TR = (0.35)(0.95)(0.9) = 0.30 \text{ mW}$$

resulting in an average beam incidance, which also equals the receiver incidance. Thus

$$E_R = 4\phi_B/\pi D^2 = 4 \times 0.30/\pi \times 2^2 = 0.095 \text{ mW/cm}^2$$

With a specular (mirror) target and without a receiver lens [Figure 7–19(a)], the incidance at the receiver plane is the same and the receiver output current is $E_T \times RE$. The responsivity of the receiver for a tungsten source is $\eta_T = 0.30$. Our source is a GaAs infrared diode radiating at 940 nm, for which $\eta_{940} = 0.80$. Therefore the responsivity has to be converted:

$$RE_{940} = RE\eta_{940}/\eta_T = 0.3 \times 0.80/0.30 = 0.80 \text{ mA/mW/cm}^2.$$

Thus the receiver output is:

$$i_R = E_R RE_{940} = 0.095 \times 0.80 = 0.076 \text{ mA} = 76 \text{ } \mu\text{A}.$$

The dark current of the phototransistor is 25 nA, and the output is well within the detectable range. However, the output current can be increased considerably using a lens at the detector and concentrating the entire beam flux to the receiver.

With the specular (mirror) reflector and receiver lens [Figure 7–19(b)], the flux into the receiver equals the reflected beam flux reduced by the angular alignment efficiency and transmission losses in the lens. Considering an angular alignment error of 10° that corresponds to the alignment efficiency from Equation 7–2 of $\eta_\alpha = \cos 10° = 0.98$ and a receiver lens having the same transmission coefficient $T = 0.95$, the flux into the receiver is:

$$\phi_R = \phi_B \eta_A T = (0.30)(0.98)(0.95) = 0.279 \text{ mW}$$

The receiver has a diameter of 0.152 cm, and the incidance at the receiver plane is

$$E_{RL} = \phi_R/A_R = 4 \times 0.279/\pi \times 0.152^2 = 15.38 \text{ mW/cm}^2.$$

The receiver current then is

$$i_{RL} = E_{RL} RE_{940} = 15.38 \times 0.80 = 12.3 \text{ mA}$$

The lens increases the receiver output by a factor of 160.

With a diffusing reflector and without a receiving lens [Figure 7–19(c)], the target sterance is, considering the reflection coefficient of the target $R = 0.80$:

$$L_T = E_T R/\pi = 4\phi_L TR/\pi^2 D^2 = (4 \times 0.35)(0.95 \times 0.80)/\pi^2 \times 2^2$$
$$= 0.0270 \text{ mW/sr/cm}^2.$$

The angular range of the target from the receiver is

$$\omega_T = A_T/d^2 = \pi D^2/4d^2 = \pi \times 2^2/4 \times 100^2 = 3.14 \times 10^{-4} \text{ sr}$$

Thus the receiver incidance is

$$E_R = L_T \omega_T = (0.0270)(3.14 \times 10^{-4}) = 0.848 \times 10^{-5} \text{ W/cm}^2$$

The corresponding receiver output is

$$i_R = RE_{940} E_R = (0.8)(0.848 \times 10^{-5}) = 0.678 \times 10^{-5} \text{ mA} = 6.78 \text{ nA}$$

This current is below the phototransistor dark current and, therefore, the system is not usable.

With a diffusing reflector and receiving lens [Figure 7–19(d)], the incidance at the lens is the same as the incidance at the receiver plane. Thus, the flux collected by the lens is:

$$\phi_{RL} = E_R A_L = (0.848 \times 10^{-5})(\pi \times 2^2)/4 = 2.66 \times 10^{-5} \text{ mW}$$

The incidance at the receiver is

$$E_{RD} = \phi_{RL} T/A_R = (2.66 \times 10^{-4})(0.95 \times 4)/\pi \times 0.152^2$$
$$= 13.9 \times 10^{-4} \text{ (mW/cm}^2)$$

and the receiver output is

$$i_{RL} = RE_{940} E_{RD} = (0.8)(13.9 \times 10^{-4}) = 11.1 \times 10^{-4} \text{ mA} = 1.11 \text{ }\mu\text{A}$$

This current is above the transistor dark current (25 nA at 25°C), but it may be difficult to detect.

The conclusion of this example is that diffusing reflectors are inefficient and can be used only when they are close to the detector.

When a reflective system is used for an application that requires high resolution or a high MTF, such as bar-code readers, another problem arises. To achieve highest possible resolution, the beam diameter should be very small and focused at the bar chart. As the operator manipulates a code reader, the distance from the reader to the chart varies. Thus, resolution is limited by the depth of field of the device. Depth of field information is usually given on the data sheet for the reader.

7–2   OPTOCOUPLERS

An optocoupler has a source that is optically coupled to a receiver. Both the source and receiver are enclosed in a sealed package. The device has two or more input terminals connected to the source and two or more output terminals connected to the output of the receiver. Despite the optical coupling, the input and output circuitry is electrically insulated, having an insulation resistance on the order of 10^{12} ohms, a capacitance less than a picofarad, and an isolation voltage strength of several kilovolts. Therefore, the device is often called an **optoisolator** since its primary purpose, besides optical coupling, is to provide good electrical insulation between the input and output circuitry.

With optocouplers, we can interconnect sensitive circuits having separate power supplies, eliminate troublesome ground current effects, and use sensitive digital circuitry to drive high-power devices. Since their appearance a few decades ago, optocouplers have become the most common control technology for high-power electrical equipment. The design principle of an optocoupler is shown in Figure 7–20.

The **optocoupler** has two separate circuits. The input circuit contains a radiation source. The flux from that source is coupled to a detector that is part of the output circuitry, here called the **receiver.** The receiver may have two or more output terminals. As mentioned, the main feature of this device is that the source and receiver are optically coupled but electrically insulated.

Classification and Specifications

Optocouplers can be classified by source characteristics, receiver circuitry and function, and general specifications. **Optocoupler general specifications** are parameters common to all optocouplers. They describe the most basic functions of a coupler. They are:

a. operating speed
b. forward current transfer ratio
c. isolation
d. limiting parameters and aging characteristics

FIGURE 7–20   Basic optocoupler.

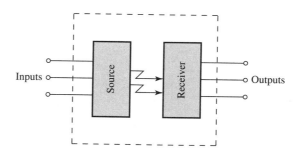

Operating speed is the primary concern with optocouplers used for digital data transmission. Several methods are used to specify the speed. The most common are **bandwidth** and **baud (Bd) rate.** The bandwidth specification depends on the pulse modulation method used. With the most efficient method, **NRZ format** (consecutive HIGH is not returned to zero or Non Return to Zero), the bandwidth is

$$f_B = 1/t_{BIT} \qquad\qquad (7\text{--}23)$$

where f_B = NRZ bandwidth (Hz)
t_{BIT} = time duration of the shortest code symbol (Bit) (s).

The **baud rate** is the greatest number of ones or zeros the system can transmit in one second. Since one bit may consist of a 1 and a 0, the baud rate is twice the NRZ bandwidth.

Another method of specifying speed is by the rise and fall times of the output signal. In the case of optocouplers, the output waveform not only has longer rise and fall times than the input, but the entire signal is delayed with respect to the input signal. This effect is called **propagation delay** or **propagation time,** defined with the help of Figure 7–21. These parameters are not constant for the device but depend considerably on the driving current and driving circuitry. The longer the propagation delay, the longer the rise and fall times, and the narrower the NRZ bandwidth.

The **forward current transfer ratio, (CTR),** is the ratio of the optocoupler output current to the input current, often expressed in percent:

$$CTR = (I_0/I_I) \times 100 \qquad\qquad (7\text{--}24)$$

where CTR = current transfer ratio (%)
I_0 = optocoupler output current (A)
I_I = optocoupler input current (A).

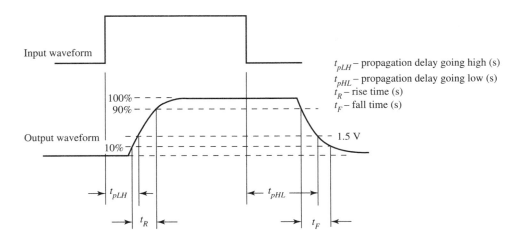

FIGURE 7–21   Definition of rise-and-fall time and propagation delay.

Depending on the optocoupler, the *CTR* may have a value of several percent to thousands of percents. This parameter also varies with input current, ambient temperature, and the external circuit configuration.

The input of an optocoupler often carries two signals: the desired input signal to be transferred to the output and an undesired stray signal that should be isolated from the output. **Isolation** describes the optoinsulator's ability to discriminate between the desired and stray input signals. Often the term **differential mode signal** is used for the desired signal and **common mode signal** for the undesired signal. The common mode signal may originate from different ground potentials for the input and output circuitry, capacitively coupled voltages to the input circuitry, magnetic pulses received through inductive pickups, and many other ways. Quite often the common mode signal voltage exceeds by many times the differential mode signal voltage. A typical condition with differential and common mode voltages is shown in Figure 7–22.

The common mode voltage can affect the output voltage in two ways: by modulating the input current i_D and by affecting the output current through stray capacitances in the optocoupler. Input current modulation depends on input circuitry and how the common mode voltage is introduced into the circuit. It can be neutralized somewhat by grounding the input circuit through a potentiometer, as shown in Figure 7–22.

Common voltage effect through internal coupling within the optocoupler is beyond the user's control. It can be reduced by selecting an optocoupler with an internal electrostatic shield. This option reduces the coupling capacitor (C_{CM}) value from a typical 0.07 pF to 0.007 pF. Also, when an optocoupler with an active transistor in the output circuit is used, neutralization of the common mode coupling is possible. For this, a neutralizing capacitor

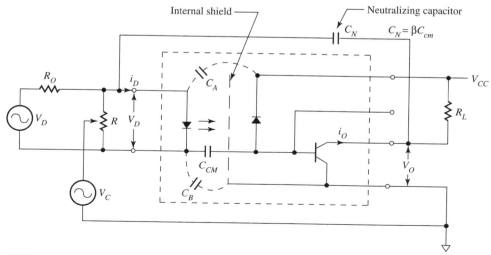

FIGURE 7–22   Differential and common mode voltages with an optocoupler.

with a value $C_N = \beta C_{CM}$, where β is the gain of the output transistor, should be used. With $C_{CM} = 0.07$ pF and a typical $\beta = 100$, the value of the neutralizing capacitor is 7 pF. The capacitor should have the same voltage rating as the optocoupler.

The optocoupler's ability to suppress a common mode signal is expressed in the **common mode rejection ratio (CMRR),** defined as follows:

$$CMRR = VTR_D/VTR_C \qquad \textbf{(7–25)}$$

where $CMRR$ = common mode rejection ratio

VTR_D = voltage transfer ratio for differential mode signal; $VTR_D = dV_0/dV_D$

VTR_C = voltage transfer ratio for common mode signal; $VTR_C = dV_0/dV_C$.

Often the same term is expressed as **common mode rejection (CMR),** which is the logarithm of $CMRR$ expressed in decibels:

$$CMR = 20 \log(CMRR) \qquad \textbf{(7–26)}$$

where CMR = common mode rejection (dB).

Besides common mode rejection, the **insulation** between input and output circuits is an important parameter. The insulation is specified as the maximum safe voltage and minimum insulation resistance between input and output circuits. The maximum safe voltage is the test voltage that the optocoupler will withstand without breakdown. It is not the safe voltage at which the device can operate for a long time. At such high voltages, a corona that deteriorates the insulation often develops.

The insulation resistance between input and output circuits is very high, 10^{12} ohms being a typical value.

Limiting parameters, such as absolute maximum power, current, and voltage ratings, are the same as for all semiconductor devices and are given in the device data sheet.

Aging characteristics are especially important for optocouplers since they are often used in applications that require 100% reliability over a long period of time. However, like any semiconductor product, they can exhibit catastrophic failures and operation deterioration during their lifetime. Catastrophic failures cannot be eliminated but can be reduced to a very low probability by selecting high reliability devices and driving them under low temperature and stress conditions.

Deterioration of CTR during the optocoupler lifetime is unavoidable. It is largely caused by a decrease of the source (LED) output and can be reduced by using modest driving conditions and by including a reserve in the initial design. Using these precautions, lifetimes in the tens of thousands hours of operation can be expected.

Source characteristics and circuit determine the optocoupler input response. Optocouplers can respond to three types of signals:

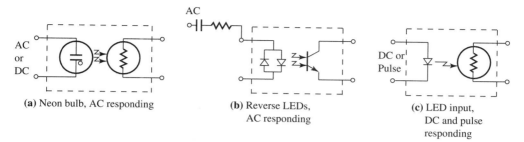

(a) Neon bulb, AC responding

(b) Reverse LEDs, AC responding

(c) LED input, DC and pulse responding

FIGURE 7–23   Basic input sources of optocouplers.

a. direct current or voltage (DC)
b. pulses, including digital signals
c. alternating current or voltage (AC)

Figure 7–23 depicts some representative examples of input sources. Figures 7–23(a) and (b) are two bipolar (AC) responding optocouplers. The neon bulb requires a high voltage (above 70 V) to activate. Both may be used to respond to an AC signal, to detect the presence of a telephone ring signal, or to alert the user to a power failure, for example.

Figure 7–23(c) shows the most common LED input optocoupler. As a diode, LED will respond to DC, unipolar pulses, and a positive half cycle of AC. The LED is usually a GaAs-based emitter with P epitaxy. It may emit wavelengths from visible to infrared. The design emphasizes close coupling to the detector, with special attention either to high speed or gain. Typical input characteristics are shown in Figure 7–24.

As indicated, the input current-voltage relationship is dependent on the ambient temperature, which is typical for an LED. To avoid overloading the device, the operating range may be divided into continuous drive or pulse

FIGURE 7–24   Typical input characteristic of an LED source optocoupler.

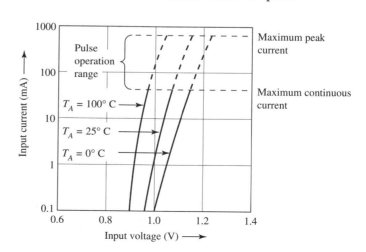

operation. Pulse operation permits a higher LED current, resulting in somewhat faster response times.

The optocouplers that are used for data transfer are connected to digital circuitry. The compatibility between the optocoupler and the digital circuit family is paramount. The couplers used with TTL circuitry should not have an input current of more than 1.6 mA. For a CMOS family, the input current should not exceed 0.5 mA. When the optocoupler input current requirement exceeds the digital circuit capability, drivers as shown in Figure 7–25 may be used.

Optocoupler Applications

Output or **receiver circuitry** determines the function and characteristics of the optocoupler. Truly, there are hundreds of optocouplers with different functions. They may be classified into six main groups:

1. optocouplers to handle digital signals
2. optocouplers to handle analog signals
3. optocouplers for control functions
4. optocouplers for AC and power switching
5. optointerrupters and optoreflectors
6. special purpose optocouplers

FIGURE 7–25 Optocoupler driving circuits.

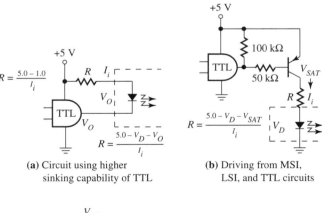

(a) Circuit using higher sinking capability of TTL

(b) Driving from MSI, LSI, and TTL circuits

(c) CMOS driving circuit

The distinction between groups is often blurred. For example, a transistor output optocoupler may handle a modest speed digital signal, has reasonable linearity for analog operation, and is often used for control functions. Special optocouplers are devices designed to meet requirements for a specific task. They are unique devices and are not described in this text.

Digital Application of Optocouplers

Optocouplers to handle digital signals require a fast response time. This is not only a function of optocoupler capability, but also of driving circuitry and the matching between the device and the transmission line that drives it. For more information on these requirements, consult the references listed under Recommended Reading at the end of this chapter.

The fastest coupler uses a fast GaAs infrared LED that is coupled to a fast PIN diode. This device, however, has low CTR and will require considerable external circuitry to operate. For this reason, it is not a very practical solution. Using integrated circuit technology, the required circuitry can be built into the optocoupler. This technique makes the device much more economical, but requires some compromises that reduce the optocoupler speed. A typical example is the HP HCPL-2400 optocoupler that operates up to a 40 MBd data rate and is compatible with the TTL, LSTTL, and HCMOS families. Figure 7–26(a) shows the internal circuit of such a device, consisting of a pulse-shaping amplifier and a three-state gate. The optocoupler has propagation delay of about 30 ns, output rise time of 20 ns, and a fall time of 10 ns. Reliably operating this optocoupler requires bypassing the power supply with a 0.01 μF ceramic capacitor, located close to the supply pin.

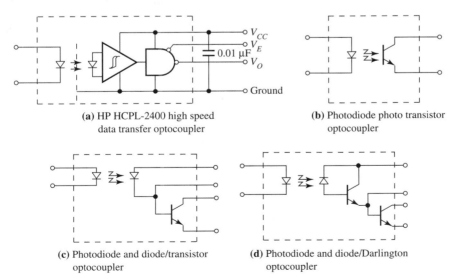

(a) HP HCPL-2400 high speed
data transfer optocoupler

(b) Photodiode photo transistor
optocoupler

(c) Photodiode and diode/transistor
optocoupler

(d) Photodiode and diode/Darlington
optocoupler

FIGURE 7–26   Typical optocouplers for data transmission.

More common receivers for optocouplers are the optotransistor, integrated optodiode and transistor, and optodiode and Darlington transistor, shown in Figures 7.26(b), (c), and (d).

The diode/transistor optocoupler is the simplest and least expensive device. However, it has the shortcomings of slow response and inherent nonlinearity. The slow response is caused by a high transistor input capacitance, due to a base collector capacitance that is transferred to the input by the Miller effect. Therefore, the response time is also dependent on the transistor gain, which, in turn, is determined by the load resistor. The typical rise and fall times are on the order of a few to tens of microseconds. The CTR of this device is typically below 100%, depending on the collector voltage. It is a relatively nonlinear function. Typical phototransistor receiver characteristics, compared with diode/transistor receivers, are depicted in Figure 7–27.

Besides having improved linearity, the diode/transistor receiver is also faster. Typical propagation delays, as shown in Figure 7–28(a), are below 1 μs and the bandwidth is about 10 MHz. The t_{PLH} is highly dependent on the load resistor and ambient temperature, as shown in Figure 7–28(b).

The current gain is not a very stable parameter. Its typical variation with input current and ambient temperature is shown in Figures 7–29(a) and (b).

Several types of output circuits can be used with the diode/transistor output. They are shown in Figure 7–30.

As shown in Figure 7–30(a), a diode/transistor can also be used as a LED-to-diode optocoupler, resulting in a faster response time and improved linearity, but very low *CTR*—on the order of 1%. Therefore, the device requires extensive external amplification. In this mode, the emitter and collector of the unused transistor should be connected to the base; otherwise distortion will occur.

More typical applications are shown in Figures 7–30(b) and (c). Circuit (b) shows a collector-coupled output, which is inverted. The advantage of

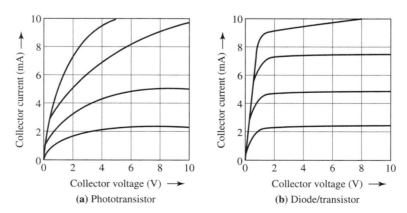

(a) Phototransistor   **(b)** Diode/transistor

FIGURE 7–27   Linearity of phototransistor and diode/transistor optocouplers.

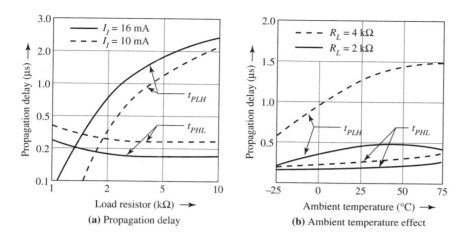

(a) Propagation delay

(b) Ambient temperature effect

FIGURE 7–28   Typical diode/transistor receiver delay characteristics.

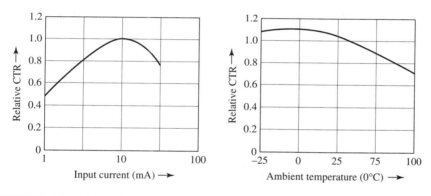

FIGURE 7–29   Typical CTR variation with input current and ambient temperature.

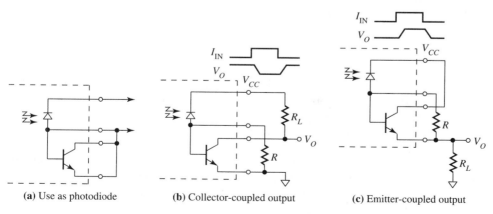

(a) Use as photodiode

(b) Collector-coupled output

(c) Emitter-coupled output

FIGURE 7–30   Optocouplers with diode/transistor output circuit.

this circuit is that it can provide considerable gain, depending on the value of the load resistor R_L. The disadvantage of the circuit is that it increases the propagation delay because of an increase of transistor input capacitance due to the Miller effect. Circuit (c) is free of the Miller effect and thus provides a faster response. When coupled to the TTL family, the gain of the circuit is limited because of the high sinking current that is required for TTL. Both circuits require a bypass resistor between the emitter and base to prevent negative bias at the base emitter junction. A typical use of the circuit is shown in Example 7–8.

EXAMPLE 7–8

Design a digital TTL family interface for a transistor output optocoupler with the following characteristics: $i_F = 16$ mA; $V_F = 1.5$ V; $CTR = 20\%$; $V_{OL1} = 0.4$ V; $V_{OL2} = 0.5$ V.

Solution
The interface circuit with TTL family data is shown in Figure 7–31.
The LED drive current is 16 mA, thus

$$i_F = (V_{CC} - V_F - V_{OL1})/R_F \quad \text{or}$$

$$R_F = (V_{CC} - V_F - V_{OL1})/i_F = (5.0 - 1.5 - 0.4)/16.0 = 0.194 \text{ k}\Omega$$

V_{OL2} is 0.5 V > 0.8 V, as required by TTL specifications. The transistor output current at ON (low) condition is:

$$i_{OL2} = i_F \times CTR = 16 \times (20/100) = 3.2 \text{ mA}$$

From the currents and voltage at node A, we can write

$$R_L = (V_{CC} - V_{OL2})/i_{OL2} + (-i_{L2}) = (5.0 - 0.5)/(3.2 - 1.6) = 2.81 \text{ k}\Omega$$

At output HI condition, the voltage drop in R_L is $V_{RH2} = R_L I_{IH2} = 2.81 \times 0.04 = 0.11$ V, which is negligible and $V_{OH2} = 5.0 - 0.11 = 4.89$ V > 2.0 V.
For reliable operation the values should be checked for component and device tolerances.

The highest CTR is available with optocouplers having a Darlington output circuit, either photodarlington or photodiode Darlington configuration, as shown in Figures 7–32(a), (b), and (c).
In principle, optocouplers with a Darlington output circuit have higher CTR (in the range of several hundred to a thousand percent) and a longer response time (on the order of tens to hundreds of microseconds). Generally, they are much more sensitive devices with the input current in the range of 1 mA and below. They are used as moderate speed line receivers for several families of digital circuits, including CMOS.

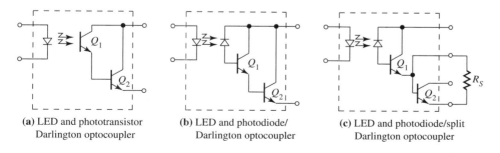

FIGURE 7–31   TTL interface circuit.

The configuration in Figure 7–32(a), with a phototransistor/Darlington, is the simplest, but has the lowest speed and poorest linearity. Configurations (b) and (c) use a photodiode as a detector, followed by a Darlington circuit. Circuit (c) has a split Darlington configuration. Its advantage is that the lowest available output voltage (output 0) equals the saturation voltage of Q_2, which is on the order of 0.1 V. In the normal Darlington circuit, the lowest available voltage is the sum of the saturation voltage of transistor Q_1 (about 0.1 V) plus the base emitter voltage drop of Q_2 (about 0.7 V), which adds 0.8 V. That voltage may be too high for some integrated circuit families. A separate Q_2 base output also allows the use of resistor R_S as a shunt from the base to the collector. This shunt effectively reduces the propagation delay t_{PLH}, but also reduces the output current and so increases t_{PHL}. However, by increasing the input current, the low value of t_{PHL} can be restored and an overall improvement is achieved. Typical Darlington-output optocoupler characteristics are shown in Figure 7–33.

When using high-speed digital optocouplers, a 0.01 μF ceramic capacitor close to the V_{CC} input pin is required to eliminate cross coupling and prevent oscillations. Using optocouplers with a longer transmission line requires special matching techniques to prevent reflections.

A typical use of a Darlington optocoupler in a medium-speed digital link is shown in Example 7–9.

(a) LED and phototransistor
Darlington optocoupler

(b) LED and photodiode/
Darlington optocoupler

(c) LED and photodiode/split
Darlington optocoupler

FIGURE 7–32   Optocoupler with Darlington-output circuit.

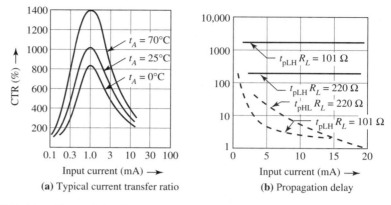

(a) Typical current transfer ratio

(b) Propagation delay

FIGURE 7–33   Typical Darlington output optocoupler characteristics.

EXAMPLE 7–9

Design a sensitive, low input current data link for the CMOS family. Use a photodarlington optocoupler and base shunt to increase data rate.

Solution
Select Darlington output optocoupler HP 6N133 with $V_{CC} = 10$ V; $i_F = 1$ mA; $V_F = 1.45$ V; CTR = 500%; $V_{OL2} = 0.1$ V. The circuit is shown in Figure 7–34. Without shunt resistor R_S, $CTR = 500\%$ and

$$R_F = (V_{CC} - V_F - V_{OL1})/i_F = (10 - 1.45 - 0.5)/1.0 = 8.05 \text{ k}\Omega$$

$$i_L = i_F \times CTR = 1.0 \times (500/100) = 5.0 \text{ mA}$$

$$R_L = (V_{CC} - V_{OL})/i_L = (10.0 - 0.1)/5.0 = 1.98 \text{ k}\Omega$$

Maximum data rate from the data sheet is 250 kHz.

With shunt resistor $R_S = 820$ Ω, the CTR is reduced to 50%. To achieve the same output, either i_F or R_L has to be increased by a factor of 10. Increasing R_L will decrease the data rate, so increasing i_F is selected. Thus,

$$R_F = (10 - 1.45 - 0.5)/10.0 = 0.80 \text{ k}\Omega$$

From the data sheet, the maximum data rate under these conditions is 650 kHz.

FIGURE 7–34   Darlington output optocoupler with CMOS interface.

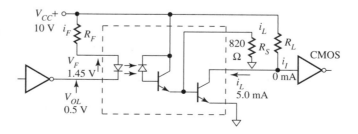

Analog Application of Optocouplers

If optocouplers are to handle **analog signals,** the primary concern is linearity. In principle, both the source (LED) and the receiver are nonlinear devices. Therefore the total transfer function, which is the product of both transfer characteristics, is also a nonlinear function. It can be described by the equation

$$I_o = k(I_{in}/I_{ref})^n \qquad\qquad (7\text{–}27)$$

where I_o = optocoupler output current (A)
  I_{in} = optocoupler input current (A)
  $k = I_o/I_{in}$ at $I_{in} = I_{ref}$
  I_{ref} = input current where k is measured (A)
  n = exponent.

For a linear device, $n = 1$. Since the source (LED) and receiver are not linear devices, the output is not a linear function of the input. Typically, n (also the slope on the log I_o/I_{in} graph) has a value about 2 at lower input levels and decreases to a value below 1 at high input levels. Representative I_o/I_{in} characteristics are shown in Figure 7–35.

  The graph shows the expected result: devices with higher CTR show worse linearity. As seen from the graph, most devices exhibit nonlinearity measured in tens of percents, even over a limited input range of 1 to 10. Therefore, this type of optocoupler is not suitable for linear applications, even when modest accuracy is required. Also, distortion is introduced when the devices are used with an input signal having AC modulation.

FIGURE 7–35   Typical transfer characteristics of various optocouplers.

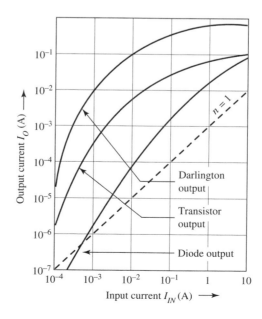

There are, however, two devices that exhibit superior linearity when compared with transistor output optocouplers: photoconductors and photo-FETs. Both have passive output circuits and act as reasonably linear variable resistors whose value can be controlled by the input current. For the photo-FET to operate in this manner, it should operate in the region below the saturation current; that is, it should operate as a passive device without a drain-source voltage. Both devices can be used as input-controlled linear AC attenuators. Typical circuits for a photoFET are shown in Figure 7–36.

Several methods are available to improve linearity. The simplest, which also reduces AC distortion, is to select the LED and output transistor operating points in such a way that nonlinearity in one is offset by opposing nonlinearity in the other, as shown in Figure 7–37.

The resistor R_1 selects the bias current I_B for the LED. Resistors R_2 and R_3 determine the operating point of the transistor. If both conditions are selected properly, the distortion of the AC signal over a 5 V swing can be kept below 1%. The values of the resistor depend on the optocoupler used and can be determined through experimentation.

A wider operating range and better linearity can be achieved with a system using servo techniques. Such a system requires two devices or a dual channel optocoupler whose gains will closely track one another over their operational range. The principle schematic is shown in Figure 7–38.

In this circuit, the LED of both channels is biased to the same operating point by bias currents I_{B1} and I_{B2}. The outputs from Q_1 and Q_2 are fed to an operational amplifier U_2. The operational amplifier tries to keep both input currents equal and therefore forces a current I_{IN2} through the second channel LED. If the gains of both channels are equal, $I_{IN1} = I_{IN2}$, and the device operates as an amplifier with a gain of one. When the gains of both channels are not equal, which is the normal condition, the offset current can be minimized with a zero adjustment potentiometer and the gain can be equalized by adjusting the resistor R_2. Under these conditions, the device exhibits stable gain as long as the gains of both channels track another. In practice, about 1% linearity over a 10 V dynamic range may be expected. The circuit works for AC

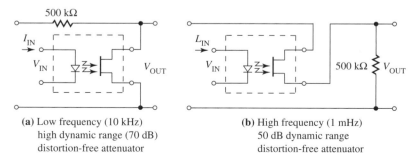

(a) Low frequency (10 kHz)
high dynamic range (70 dB)
distortion-free attenuator

(b) High frequency (1 mHz)
50 dB dynamic range
distortion-free attenuator

FIGURE 7–36   Use of photoFET output in a linear attenuator.

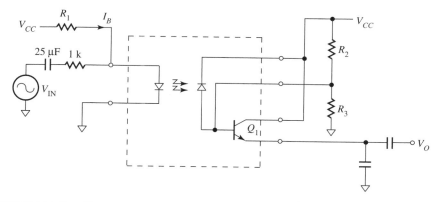

FIGURE 7–37   Circuit to linearize optocoupler characteristics.

with several tens of kilohertz bandwidth, depending on the characteristics of amplifiers U_1 and U_2.

When better linearity is required, an A-to-D converter may be used. The resulting digital signal may be transmitted through the optocoupler without any loss of linearity. Optocoupler output may be converted back to DC using a D-to-A converter. By this method, excellent linearity is achieved, but requires considerably more hardware at a higher cost. Voltage-to-frequency converters may also be used to overcome the inherent nonlinearity of the optocoupler.

Control Applications of Optocouplers

Performing **control functions** is one of the most natural and popular applications of the optocoupler. An optocoupler is an ideal device to use between

FIGURE 7–38   Servo type linear optocoupler amplifier.

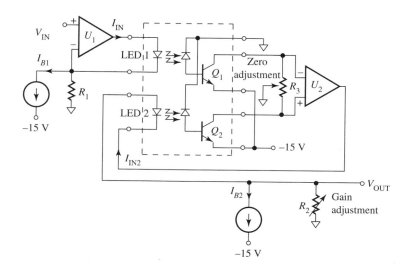

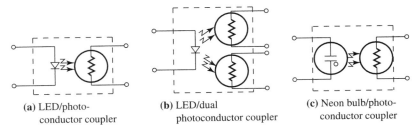

(a) LED/photo-
conductor coupler

(b) LED/dual
photoconductor coupler

(c) Neon bulb/photo-
conductor coupler

FIGURE 7–39   Typical optocouplers with photoconductor as a receiver.

delicate and sensitive digital circuitry and robust, high power level industrial circuits. This application is one where the isolation properties of the opto-coupler are most highly appreciated.

The basic control functions are turning devices ON and OFF, controlling actuators and stepping devices, and setting current and voltage level. Receivers with transistor and Darlington output circuits, described previously, are most adaptable for many of these tasks. For tasks that require control of AC-operated devices, the optocouplers with SCR and TRIAC receiver circuits are most appropriate.

Tasks such as attenuating signals and controlling voltage and current levels can be accomplished using photoconductor output optocouplers. Because of their slow response, these devices are rarely used in digital and communication fields. In control applications, however, a response time of a few

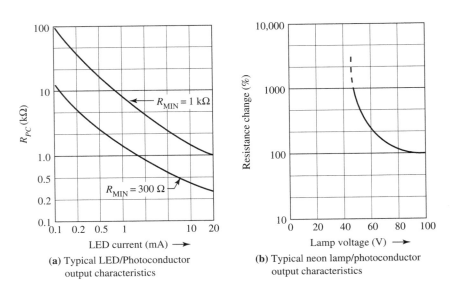

(a) Typical LED/Photoconductor
output characteristics

(b) Typical neon lamp/photoconductor
output characteristics

FIGURE 7–40   Typical photoconductor receiver output characteristics.

milliseconds is perfectly acceptable. These devices are popular in electronic musical instruments, as volume control devices and also as TRIAC drivers. A few typical examples are shown in Figure 7–39. The neon bulb input optocoupler is AC-responding and is commonly used for ring detection in telephones.

Typical output characteristics of photoconductor receiver optocouplers are shown in Figure 7–40.

As illustrated in Figure 7–40(a), the resistance of a photoconductor changes over about a two-decade range. It is not a linear function of the LED current. The neon lamp/photoconductor optocoupler exhibits a minimum of tenfold increase of output resistance when the neon bulb is OFF.

The rise and fall times of an optocoupler depend on the load resistance and driving conditions. Typical characteristics are shown in Figure 7–41.

When using this type of optocoupler as a remote controlled attenuator for volume control in musical instruments, it may cause distortion, especially at low LED current levels and high audio signal levels. A typical application is shown in Example 7–10.

EXAMPLE 7–10

Design a remote controlled stereo audio attenuator. Calculate the range of attenuation.

Solution
Use a Hamamatsu P875-G35-911 dual photoconductor output optocoupler. Its characteristics are

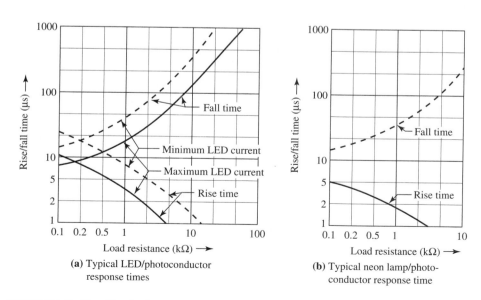

(a) Typical LED/photoconductor response times

(b) Typical neon lamp/photoconductor response time

FIGURE 7–41   Typical response time of photoconductor receiver optocoupler.

Input Current mA	LED Voltage V	Output Resistance kΩ
20	2.5	0.3
0.1	1.7	50

The circuit is in Figure 7–42.

At i_{Fmax} the value of R_1 is:

$$R_1 = (V_{CC} - V_F)/i_{Fmax} = (10.0 - 2.5)/10.0 = 0.375 \text{ k}\Omega$$

From i_{Fmin} we can find $R_1 + R_C$

$$R_1 + R_C = (V_{CC} - V_F)/i_{Fmin} = (10.0 - 2.5)/0.1 = 75 \text{ k}\Omega$$

$$R_C = 75 - 0.75 \approx 75 \text{ k}\Omega$$

Selecting $R_2 = 50 \text{ k}\Omega$, the maximum attenuation is:

$$A_{max} = R_{Pmin}/(R_2 + R_{Pmin}) = 0.30/(50 + 0.3) = 0.006 \quad \text{which is}$$

$$A_{maxdB} = |20 \log 0.006| = 44.5 \text{ dB}$$

Minimum attenuation is

$$A_{min} = R_{Pmax}/(R_2 + R_{Pmax}) = 30/(50 + 30) = 0.625 \quad \text{or}$$

$$A_{mindB} = |20 \log 0.25| = 4.1 \text{ dB}$$

The range of attenuator is $44.5 - 4.1 = 40.4$ dB.

Optocouplers to control power use photoSCR or photoTRIAC as the receiver and output devices. They often act as an interface between a controller and a power device. Relays have been used for this application for many years and are still used in this manner today. They are reliable and can

FIGURE 7–42   Remote control attenuator circuit.

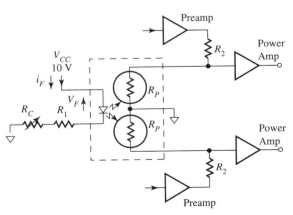

handle high power, but they have three shortcomings. First, they require large drive power and, thus, are not compatible with digital circuits. Second, the isolation between the power and control circuit is marginal for digital circuitry. And third, the coil in the relay is an inductor, which introduces high and dangerous switching spikes. Switching optocouplers are free of these shortcomings. They are compatible with digital circuits and provide high isolation. Their only shortcoming is that they have limited power capability and may require an additional power switch, a high power TRIAC, or even a relay or contactor for high power applications.

Two basic types of power switching optocouplers are available. They are shown in Figure 7-43.

Figure 7-43(a) shows a photoSCR receiver optocoupler that is suitable for switching DC current. Figure 7-43(b) shows a photoTRIAC receiver suitable for AC power switching. Both devices act as latching relays—once triggered by the LED, they will stay ON while power to the output is applied. Thus they are two-state devices, in contrast to the optocouplers described earlier that show an output response proportional to the input signal. The OFF and ON states are shown in Figures 7-44(a) and (b).

A most important consideration when applying optocouplers with a photoSCR or photoTRIAC receiver is to use sufficient input current to turn the output to an ON state. The required input current for this condition depends on many factors: temperature, anode-cathode voltage, cathode gate resistance, and pulse width and duration when a pulse trigger is used. Thus, the triggering range of the input current is wide, as shown in Figure 7-45. The manufacturer's data sheet should be consulted to guarantee triggering under all possible conditions.

FIGURE 7-43   Common power switching optocouplers.

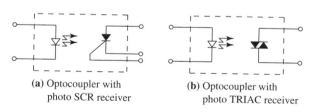

(a) Optocoupler with photo SCR receiver

(b) Optocoupler with photo TRIAC receiver

FIGURE 7-44   Output characteristic of SCR and TRIAC output optocoupler.

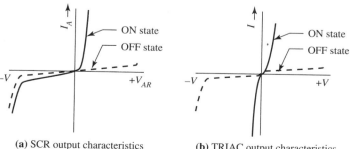

(a) SCR output characteristics

(b) TRIAC output characteristics

FIGURE 7–45 Typical range of required input current for SCR and TRIAC triggers.

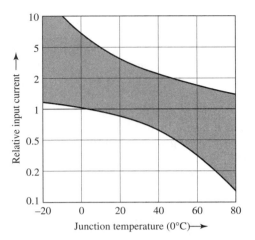

All thyristors can be triggered by a high output voltage gradient or an output voltage spike. This situation can be avoided by installing a low pass filter at the optocoupler output terminal, as shown in Figure 7–48.

The delay in a typical thyristor receiver optocoupler is on the order of a few microseconds. Thus, the device is fast enough to switch during any fraction of the phase of the applied signal and can be used to control the *RMS* power of the signal.

A few typical SCR receiver optocouplers are illustrated in Figure 7–46. Figure 7–46(a) shows the basic circuit of a coupler acting as a DC switch or latching relay. The value of the gate cathode resistance R_1 depends on the type of SCR, input current, and load voltage. Information on these is available on the optocoupler data sheets. As with all semiconductor devices, maximum steady and peak current and reverse voltage should be considered. When the device is used to switch incandescent loads (lamps) or motors, inrush or starting current of the device should be considered. These currents may be 10 to 15 times higher than the operating current and should not exceed the maximum allowable peak current of the device. A capacitor C_1 across R_1 helps to prevent or suppress voltage transients in the load circuit and avoid undesired triggering.

When a higher load has to be switched, an optocoupler can be used to trigger a high power SCR with a proper rating, as shown in Figure 7–46(b). This circuit contains an $R_2 C_1$ **snubber circuit** that also suppresses the transients in the load circuit. Component values are usually given in the data sheet.

The SCR optocoupler can also be used to trigger a TRIAC and thus control AC power, as shown in Figure 7–46(c). This use requires an additional rectifier bridge.

Many other variations, such as normally closed relays, and zero voltage switching relays, are available. Also, DC switching can be accomplished with

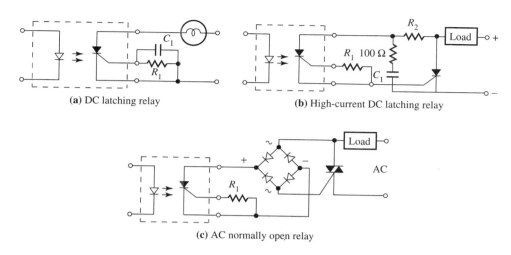

(a) DC latching relay

(b) High-current DC latching relay

(c) AC normally open relay

FIGURE 7–46   Typical circuits with SCR receiver optocouplers.

a transistor receiver optocoupler and an external SCR. A typical design is illustrated in Example 7–11.

EXAMPLE 7–11

Design a TTL logic driven latching relay to activate a 0.1 A, 12 V DC lamp.

Solution

A General Instrument MCS21 optocoupler with SCR output is selected for the task. Its specifications are $i_F = 20$ mA; $V_F = 1.2$ V; $R_G = 27$ kΩ; $C_G = 0.01$ μF. The circuit is shown in Figure 7–47.

From input condition

$$R_F = (V_{CC} - V_F - V_{OL})/i_F = (5.0 - 1.2 - 0.4)/16 = 0.212 \text{ k}\Omega$$

The maximum optocoupler peak output current is 1.0 A, sufficient for the lamp's inrush current. R_G and C_G are selected from the data sheet to suppress transients up to 20 V with gradient up to 1000 V/μs.

For AC switching, an optocoupler with a TRIAC receiver provides the simplest solution. Basic circuits are shown in Figure 7–48.

FIGURE 7–47   DC latching relay circuit.

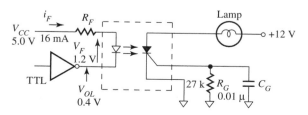

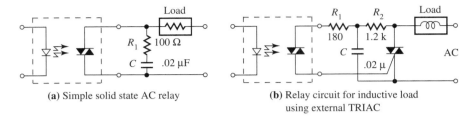

(a) Simple solid state AC relay   (b) Relay circuit for inductive load
using external TRIAC

FIGURE 7–48   Typical circuits with TRIAC receiver optocouplers.

The simplest circuit, Figure 7–48(a), uses the optocoupler TRIAC directly for switching. This option, however, is limited to small loads (below 1 A) and for sources that are free of transients. The components R and C provide only limited transient protection.

For most conditions, the optocoupler is used to trigger an external TRIAC that is rated according to the load requirements, as in Figure 7–48(b). The snubber circuit, consisting of components R_1, R_2, and C_1, should be selected according to the gate requirements of the external TRIAC. The values given in the schematic are typical for a nonsensitive gate TRIAC. This circuit is usable for inductive loads as well.

Example 7–12 illustrates a typical design employing a TRIAC output optocoupler.

EXAMPLE 7–12

Design a TTL logic controlled switch for a 100 W 120 V AC lamp and inductive (motor) load.

Solution
Use a TRIAC output optocoupler to trigger a power switching TRIAC. Select optocoupler Motorola MOC 310: $i_F = 8$ mA; $V_F = 1.2$ V. TRIAC GE SCI368: $i = 3$ A: $I_{PEAK} = 30$ A; V = 200 V AC. The circuit is shown in Figure 7–49.

A restive load circuit is shown in Figure 7–49(a). From the input condition

$$R_F = (V_{CC} - V_F - V_{IL})/i_F = (5.0 - 1.2 - 0.4)/8.0 = 0.425 \text{ k}\Omega$$

From the data sheet a 180 Ω gate resistance is required. TRIAC peak current of 30 A is sufficient for lamp inrush current.

For an inductive load a snubber circuit, as shown in Figure 7–49(b), is required. The values for the components are taken from the data sheet.

There are many types of TRIAC-optocoupler applications. Further information can be found in any standard TRIAC handbook. Most circuits may be used by just changing the trigger circuit to conform with optocoupler output.

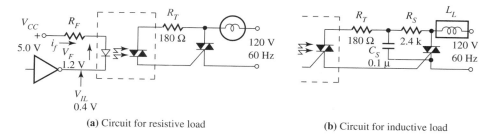

(a) Circuit for resistive load   (b) Circuit for inductive load

FIGURE 7-49   AC switching circuit.

Optointerrupters and **optoreflectors** are optocouplers in which source and detector are separated in such a way that the flux from the source to the detector can be blocked by a mechanical object. An **optointerrupter** has a 0.2 to 1 mm wide slot between the source and the detector where a thin sheet of paper or metal can be fitted. The device can be used as a simple and inexpensive edge detector, disk encoder, or punch-card reader.

In an **optoreflector** the source has a lens that focuses the flux to a distance of a few millimeters. When a reflective surface is placed there, flux is reflected back to the detector. The device may be used as an edge or tape end detector, and many other similar purposes. Pictures of a typical optointerrupter and optoreflector are shown in Figure 7-50.

The application of optointerrupters and optoreflectors is well described in manufacturers' catalogs. A typical use is shown in Example 7-13.

EXAMPLE 7-13

Design an edge detector using an optointerrupter and optoreflector. Show resolution of the detectors.

Solution a
A Hamamatsu P3784 optointerrupter is chosen for the task. Its specifications are $i_F = 5$ mA; $V_F = 1.1$ V; $i_C = 1.5$ mA. The CMOS logic family is selected for control. The circuit is shown in Figure 7-50(a).

The value of R_F is:

$$R_F = (V_{CC} - V_F)/i_F = (5.0 - 1.1)/5.0 = 0.780 \text{ k}\Omega$$

For CMOS control circuit with practically no input current, the value of R_L is:

$$R_F = (V_{CC} - V_{OL})/i_C = = (5.0 - 0.4)/1.50 = 3.07 \text{ k}\Omega$$

From the interrupter transfer characteristic it can be seen that output is switched over 0.2 mm movement of the edge. Thus the resolution of the interrupter is very high.

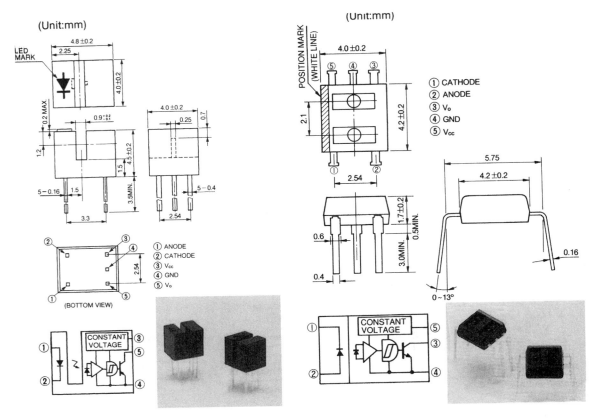

FIGURE 7–50   Typical optointerrupter and optoreflector. (Courtesy Hamamatsu Corporation.)

Solution b

A high resolution optoreflector Hamamatsu P3062-01 is selected. Its specifications are $i_F = 20$ mA; $V_F = 1.3$ V; $V_{OL} = 0.1$ V; $V_{OH} = V_{CC} = 5.0$ V: $R_L = 280\ \Omega$. The device has an internal Smitt trigger for very rapid output switching. The circuit is shown in Figure 7–50(b).

The value of R_F is:

$$R_F = (V_{CC} - V_F)/i_F = (5.0 - 1.3)/20 = 0.185\ \text{k}\Omega$$

As seen from the transfer characteristic, the switching happens over a very short movement, less than 0.1 mm. The hysteresis from approaching from two directions is less than 0.2 mm.

As seen from this example, optointerrupters and optoreflectors are simple, high resolution and cost effective edge detectors.

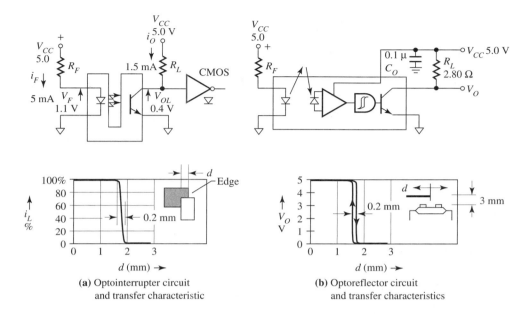

FIGURE 7–51   Edge detection using an optointerrupter or optoreflector.

SUMMARY

Chapter 7 discusses optical sensors and optocouplers. The principle of sensing is explained and terms for source detector alignment, wavelength matching, and coupling efficiency are introduced. Several methods of edge and bar detection are explained, and the term, modulation transfer function, is defined. Use of moiré patterns is demonstrated. Further, transmissive and reflective sensing systems using lenses are discussed and methods of their design and optical transfer function calculations are demonstrated.

   The operating principles of optocouplers are discussed, and main parameters, operating speed, forward current transfer ratio, insulation, common mode rejection ratio, and limiting parameters and aging characteristics are defined. Further, source and receiver (output) characteristics of the optocouplers are discussed. Couplers are divided into four groups by output characteristics: couplers for digital signals, those for analog signals, and those used for control and those used for power switching. Characteristics and design principles for each group are given. An example using optointerrupter and optoreflector is given.

RECOMMENDED READING

Hewlett-Packard. 1981. *Optoelectronics Fiber-Optics Applications Manual.* New York: McGraw-Hill Book Company.

Hewlett-Packard. 1993. *Optoelectronics Designer Catalog.* Hewlett-Packard Corp.

Whyatt, C.L. 1991. *Electro-optical System Design.* New York: McGraw-Hill Book Company.

Capell, A. 1978. *Optoelectronics: Theory and Practice.* New York: McGraw-Hill Book Company.

PROBLEMS

1. Find alignment efficiency for
   a. a flux cone of 5 mm diameter that is directed to a 3 mm diameter detector
   b. a 3 mm diameter flux cone that is directed to a 3 mm diameter detector with 1.5 mm alignment error.
2. Calculate the wavelength alignment efficiency for a source and detector with hypothetical spectral characteristics as in Figure 7–52.
3. Calculate the spacial frequency of a pattern with bar ant space width of 1/128 inch.
4. A confocal lens system couples a 750 nm Lambertian source of $I_o = 100$ mW/sr to a detector. Two lenses of 15 mm diameter are used and the first lens is placed 40 mm from the source. Use Figure 2–14 for transmission efficiency. Find OTF and flux received by the receiver.

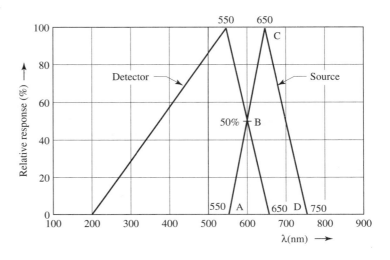

FIGURE 7–52 Hypothetical source and detector wavelength response.

5. A 10 mm diameter source with luminance of 500 cd/m² is illuminating a 5 mm diameter photodiode 20 cm away. The diode responsivity is 0.5 μA/lx. Find diode photocurrent when

   a. no lens between the source and diode is used

   b. a *F-NUM* = 2 lens that focuses the source image to the photodiode is used.

6. In a burglar alarm system the flux from an LED with I_o = 20 mW/sr and $I_\theta = I_o \cos^3 I_o$ is focused using a 1 inch diameter lens 2 inches away to a mirror 12 ft away. The reflected beam with diameter of 2 inches is directed by a 1 inch diameter lens to a detector with *RE* = 0.6 A/W. Calculate the detector current when lens T_L = 0.85 and mirror T_M = 0.8.

7. What is the *CTR* for

   a. a photodiode output optocoupler with I_I = 10 mA and I_o = 0.3 mA and

   b. a photodarlington output optocoupler with I_I = 2 mA and I_o = 12 mA?

8. At a given optocoupler circuit 5.0 V applied to input produces a 12 V output signal. A 150 V common mode signal produces 0.2 V output signal. Calculate CMRR and CMR.

9. Design a TTL input, CMOS output circuit for an optocoupler with input characteristics: i_F = 10 mA; V_F = 1.2 V; V_{CC} = 5.0 V. Output characteristics: *CTR* = 200%; V_{OL} = 0.2 V; V_{CC} = 10 V.

10. A photoconductor output optocoupler has characteristics as shown in Figure 7–40 upper curve. Design an input circuit (fixed resistor in series with variable resistor) to change the output resistance value from 1 kΩ to 10 kΩ. Input condition: V_{CC} = 10 V; V_F = 1.5 V.

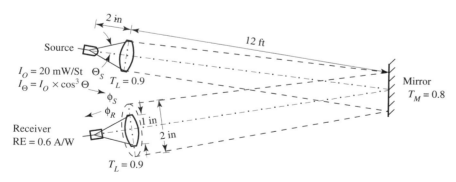

FIGURE 7–53 Burglar alarm system of Problem 7-6.

8
Principles of Fiber Optics

Of all optoelectronics developments, fiber optics has the greatest potential to impact our lives. We are entering an era in which instant information interchange is a predominant force shaping our society. Fiber optics, transmission of light signals along transparent fiber, is the key to this technology. Before the end of the century, most households will have a fiber optics link providing access not only to telephone and multichannel TV service, but also to an unlimited computer information base.

Compared with other communication methods (copper wires, coaxial cables, microwave links, etc.), fiber optics has the widest bandwidth, is the most reliable, and is cost competitive. Fiber optics is the only method free from external disturbances. A single, modern optoelectronics cable across the oceans could supply all the communication links needed between the continents.

Fiber optics applications, however, are not limited to communication. Imaging and retrieving images from normally inaccessible locations with fiber optics is widely used in medicine. Fiber optics sensors can measure displacement, temperature, magnetic field, and many other physical parameters. This technology is new and rapidly expanding.

The basic principles of fiber optics communications are covered in this text. These principles should enable the reader to design short-length communication links between devices and provide a good foundation for further study in more advanced texts (see the recommended reading list at the end of this chapter).

8–1   FIBER OPTIC FUNDAMENTALS

Chapter 2 describes how, when a light ray passes from a denser medium to less dense media, it bends closer to the media interface. Figure 8–1(a) shows that when the approaching ray angle θ_1 exceeds the **critical angle** θ_c, complete internal reflection takes place and the approaching ray cannot escape to medium 2; it is, then, completely reflected back to medium 1.

The critical angle is

$$\theta_c = \arcsin(n_2/n_1) \tag{8–1}$$

where θ_c = critical angle (°)
 n_1 = refractive index of medium 1
 n_2 = refractive index of medium 2.

When this condition is applied to a cylindrical transparent fiber, as shown in Figure 8–1(b), certain rays that approach the fiber under an angle that is less than the **acceptance angle** θ_A are completely captured by the fiber and cannot escape. Thus, the fiber performs as an **optical waveguide** or **lightguide.** Any rays that approach the fiber at an angle greater than θ_A escape from the fiber.

The acceptance angle determines the entrance condition of the fiber, and we can assign a numerical aperture value to it:

$$NA_A = \sin \theta_A = n_2/n_1 \tag{8–2}$$

where NA_A = entrance numerical aperture of the fiber.

To capture the entire flux from a source, the source flux angle or numerical aperture must be smaller than the fiber acceptance angle or the numerical aperture.

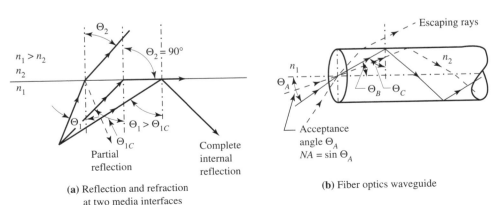

(a) Reflection and refraction
at two media interfaces

(b) Fiber optics waveguide

FIGURE 8–1   Complete internal reflection in a fiber optics lightguide.

EXAMPLE 8–1

Calculate the acceptance angle and numeric aperture for a fiber with refractive index $n_1 = 1.53$. The fiber is used in air with $n_2 = 1.00$.

Solution

From Equation 8–1, the acceptance angle is

$$\theta_A = \arcsin(n_2/n_1) = \arcsin(1.00/1.53) = 40.8°$$

From Equation 8–2, the numerical aperture is

$$NA = \sin\theta_A = 0.653$$

The above fiber optics waveguide is a good example of the basic operation of an optical waveguide. In practice, this type of fiber is not usable for two reasons. First, its surface condition is hard to control, especially when the fiber comes in contact with surrounding material. As a result, a significant fraction of the fiber flux may escape. Second, in the case of total internal reflection, an electromagnetic disturbance, called an **evanescent wave,** is generated at the interface, disturbing the propagation of the light wave. This type of fiber has very high losses and can be used only over a short distance.

Both problems can be corrected by covering the fiber or **core** with a **cladding,** as shown in Figure 8–2.

When the cladding has a lower refractive index than the core (that is, $n_1 > n_2$), complete internal reflection at the core-cladding interface occurs, and a usable lightguide is formed. The acceptance angle and numerical aperture of this type of lightguide can be calculated, assuming that the fiber is used in air ($n_0 = 1.00$):

$$\theta_A = \arcsin\sqrt{n_1^2 - n_2^2} \tag{8–3}$$

$$NA = \sin\theta_A = \sqrt{n_1^2 - n_2^2} \tag{8–4}$$

where n_1 = refractive index of core
n_2 = refractive index of cladding.

FIGURE 8–2   Gladed fiber optics lightguide.

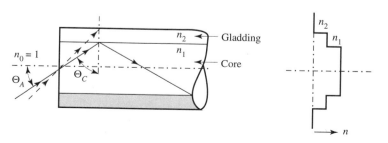

EXAMPLE 8–2

Find the acceptance angle and numerical aperture of a cladded lightguide with $n_1 = 1.53$, $n_2 = 1.50$.

Solution
From equations 8–3 and 8–4, we have

$$\theta_A = \arcsin\sqrt{1.53^2 - 1.50^2} = 17.5°$$

and

$$NA = \sqrt{n_1^2 - n_2^2} = 0.301$$

Because of great reduction of losses, all fiber optics lightguides are cladded. It is also clear from the above examples that the cladded fibers have smaller acceptance angles and numeric apertures. This property makes coupling to a cladded waveguide less efficient. In the case of a source with a Lambertian intensity pattern, where coupling to the receiver is proportional to the square of numerical aperture, the coupling in Example 8–1 is $\phi_S/\phi_R = 0.653^2 = 0.426$. In the case of the cladded fiber in Example 8–2, the coupling is $\phi_S/\phi_R = 0.301^2 = 0.096$. Over four times more flux from a Lambertian source is coupled to the uncladded fiber. This loss of flux, however, is offset by the lower losses in the cladded lightguide. This type of cladded fiber is called a **step index fiber.**

8–2 CHARACTERISTICS OF FIBERS

Most fiber optics lightguides are used for transmitting information over long distances. Thus, the important characteristics of the fiber are wide information bandwidth and low loss.

Fiber bandwidth is determined by an effect called **dispersion.** Dispersion is the phenomenon wherein light rays entering the fiber at different approach angles have different travel paths in the fiber, as shown in Figure 8–3. Here ray 1 is an **axial ray** and has the shortest path through the fiber. Ray 2 is a **meridional ray** and its path is longer. Ray 3 is a **skewed ray** that is not in a plane of central axis. Its path is also longer.

Due to different path lengths, as in a step index fiber, the travel time of these rays differs—they do not exit the end of the cable at the same time. As a result, when the cable is excited with a narrow pulse, the pulse energy spreads, and it exits the cable wider than before, as shown in Figure 8–3(b). This effect limits the pulse rate that such a fiber can handle and thus affects the fiber bandwidth.

Fibers with several possible ray paths are called **multi-mode** fibers. In these fibers, complex mode patterns similar to the patterns in a microwave

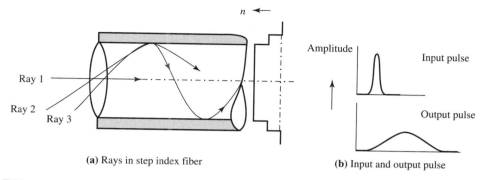

(a) Rays in step index fiber   (b) Input and output pulse

FIGURE 8–3   Ray path and pulse dispersion in a step index fiber.

waveguide develop, causing a large pulse dispersion and a limited signal bandwidth. This phenomenon is called **modal dispersion.**

Chromatic dispersion can also occur in a multi-mode fiber when rays of different wavelength enter the fiber. Since the refractive index or propagation velocity depends on the wavelength, different wavelengths have different propagation times, and pulse dispersion occurs.

The approximate time delay between different modes, or dispersion, can be calculated using the following equation:

$$\tau_S = n_1(n_1 - n_2)l/n_2c \tag{8–5}$$

where τ_S = multipath time dispersion (s)
   l = fiber length (m)
   n_1 = refraction index of the core
   n_2 = refraction index of the cladding
   c = velocity of light 3.00×10^8 m/s.

In cladded fibers $n_1 \approx n_2$, and the equation simplifies to:

$$\tau_S = (n_1 - n_2)l/c \tag{8–6}$$

The relationship among dispersion time, **maximum bit rate,** and **signal bandwidth** is:

$$B \approx 2\Delta f \approx 1/\tau \tag{8–7}$$

where B = maximum bit rate (bits/s)
   Δf = maximum signal bandwidth (Hz).

Example 8–3 illustrates the magnitude of dispersion in uncladded and cladded fibers.

EXAMPLE 8–3

Calculate time dispersion, maximum bit rate, and signal bandwidth for a cable 1 km long, using the fibers in Examples 8–1 and 8–2.

Solution
Uncladded fiber $n_1 = 1.53$, $n_2 = 1.00$.

$$\tau = n_1(n_1 - n_2)l/n_2c = (1.53)(1.53 - 1.00)(10^3/1.00)(3.00 \times 10^8)$$

$$= 2.70 \times 10^{-6} \text{ s} = 2.70 \text{ } \mu s$$

$$B = 1/\tau = 1/2.70 \times 10^{-6} = 0.370 \times 10^6 \quad \text{and} \quad \Delta f = 0.370 \times 10^6/2$$

$$= 185 \times 10^3 \text{ Hz}$$

The bit rate is only 370 kB and the bandwidth is 185 kHz.
For the cladded fiber $n_1 = 1.53$, $n_2 = 1.50$ and

$$\tau = (n_1 - n_2)l/c = (1.53 - 1.50) \times 10^3/(3 \times 10^8) = 0.10 \times 10^{-6}$$

$$= 0.10 \text{ } \mu s$$

$$B = 1/\tau = 1/(0.10 \times 10^{-6}) = 10.0 \times 10^6 \quad \text{and} \quad \Delta f = 10.0 \times 10^6/2$$

$$= 5.0 \text{ MHz}$$

The bit rate of the cladded fiber is 10 MB and the bandwidth is 5 MHz.

As seen from Example 8–3, the cladded fiber is vastly superior to the uncladded fiber with respect to signal bandwidth. However, the cladded fiber bandwidth is not nearly as wide as we would expect from a light frequency carrier. Dispersion limits the bandwidth. Fortunately, there is a type of fiber, the **graded index fiber,** that greatly reduces dispersion.

In contrast to step index fiber, the core of graded index fiber has a variable index with highest value at the central axis. The value drops as we travel away from the center, according to a parabolic law, as seen in Figure 8–4. Figure 8–4(a) shows the index profile in a step index fiber, Figure 8–4(b) shows the index profile in a graded index fiber, and Figure 8–4(c) illustrates a typical ray path in a step index fiber lightguide. The bending of the ray in a graded index fiber takes place not at the core-cladding interface (as in a step index fiber), but within the core. As a result, the higher order modes of propagation are suppressed and the dispersion is reduced. In addition, the meridional rays that have a longer traveling path than the axial ray travel in a media with a lower refractive index and have a higher velocity. This property also equalizes the traveling time and reduces the dispersion. When the graded index profile is optimized for lowest dispersion, it can be calculated by using

$$\tau_G = (n_1 - n_2)^2 l/8n_1c \qquad \qquad \textbf{(8–8)}$$

The quantities here are the same as in Equation 8–6. When comparing these two equations, we find that the dispersion in a single mode fiber is reduced by the factor of

$$\tau_G/\tau_S = (n_1 - n_2)/8n_1 \qquad \qquad \textbf{(8–9)}$$

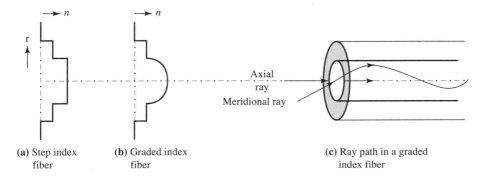

(a) Step index   (b) Graded index   (c) Ray path in a graded
   fiber   fiber   index fiber

FIGURE 8–4   Index profile and ray path in graded index fiber.

EXAMPLE 8–4

Calculate time dispersion, maximum bit rate, and signal bandwidth for a graded index fiber with $n_1 = 1.53$, $n_2 = 1.50$, and $l = 1.00$ km (as in Example 8–3).

Solution

$$\tau_G = (n_1 - n_2)^2 l/8n_1 c = (1.53 - 1.50)^2 \times 10^3/8 \times 1.53 \times 10^8$$

$$= 0.74 \times 10^{-9} \text{ s} = 0.74 \text{ ns}$$

$$B = 1/\tau_G = 1/0.74 \times 10^{-9} = 1.35 \times 10^9 = 1.35 \text{ GB/s}$$

$$\Delta f = 1.35 \times 10^9/2 = 0.77 \times 10^9 = 0.77 \text{ GHz.}$$

The graded index fiber shows improvement of about 135 times over the step index fiber.

A common graded index fiber with 50/125 μm core/cladding diameter exhibits dispersion on the order of 10 to 100 ps/km and a bit rate of 10 to 100 Gbits/skm. This performance is a remarkable improvement over the step index fiber.

Reducing the fiber diameter further improves graded index fiber performance. When the fiber diameter approximates the wavelength being used (typically about 10 μm with a 125 μm cladding diameter), the meridian rays and the higher modes of propagation are further reduced, diminishing the dispersion to 3 to 6 ps/km and resulting in a bit rate of approximately 200 Gbits/skm. This type of fiber is called a **single mode fiber.** Since chromatic dispersion still exists in all types of fibers, a further improvement in signal bandwidth can be achieved by using monochromatic sources, for example, lasers instead of LEDs.

In addition to step index and graded index profiles, several other profiles and variations of these two are used to compress higher propagation modes in the fiber.

When discussing communication bandwidth, one should not overlook the **soliton pulse** which can travel in a single mode fiber with practically no dispersion. With soliton pulses, a pulse rate of 2.9 terabit/skm (1 terabit = 10^{12} bits) and long-haul rates of 5 Gbits/s over 10,000 km have been achieved.

The soliton pulse is based on the principle of the **optical Kerr effect.** When the intensity of light exceeds a certain level, the same wavelength can propagate at different speeds, depending on the intensity. If a pulse, usually in the picosecond range, is generated so that each wavelength component has a specific amplitude traveling at the same speed, it can travel without dispersion. This condition exists until the pulse amplitude drops below the required level, after which the soliton slowly decays. When the amplitude is kept constant using modern fiber amplifiers—called **erbium-doped fiber amplifiers**—a soliton can travel very long distances without dispersion.

A short note about erbium-doped fiber amplifiers that are just now making their debut. This amplifier is a laser-like device built directly into the fiber. By doping the fiber core with the rare earth erbium and pumping it using laser diodes, an atomic multi-energy level system is generated in the core. Under these conditions, a simulated emission triggered by core flux or photons results in direct amplification of the optical signal. Using this type of amplifier and soliton pulses, data rates over 100,000 GBit/km per second have been achieved.

Fiber losses or **attenuation** determine the repeater spacing in a fiber communication link and greatly affect its cost. The attenuation is measured in dB/km.

Three intrinsic mechanisms cause the signal in a fiber to attenuate. For glass or silica fibers, the components of attenuations are as follows.

(1) *Absorption due to interaction of atomic electrons with the radiation.* Absorption peaks in the ultraviolet region and decreases rapidly toward infrared. It can be described using an empiric formula that is valid for silica fibers:

$$A_A = A_0 \exp(\lambda_0/\lambda) \tag{8–10}$$

where A_A = attenuation due to UV absorption at a wavelength λ (dB/km)
    A_0 = constant A_0 = 1.108 dB/km
    λ_0 = constant λ_0 = 4.58 μm
    λ = wavelength (μm).

(2) *Absorption due to interaction with molecular vibration.* This is a loss with peak absorption in the infrared region (approximately 10 μm for silica). It can be described as

$$A_I = B_0 \exp(-\lambda_I/\lambda) \tag{8–11}$$

where A_I = attenuation due to interaction at a wavelength λ (dB/km)
    B_0 = constant B_0 = 4×10^{11} dB/km
    λ_I = constant λ_I = 48 μm
    λ = wavelength (μm).

This component peaks at infrared and diminishes rapidly toward ultraviolet.

(3) **Rayleigh scattering.** This scattering is caused by composition fluctuation of thermodynamic origin of the glass fiber material. The size of these fluctuations is smaller than the wavelength of light. The mechanism of this attenuation is different from the other two, which cause absorption or change the light energy to heat. Rayleigh scattering, on the other hand, causes the light energy to escape from the fiber. The attenuation due to Rayleigh scattering can be described by an empirical equation:

$$A_R = C_0/\lambda^4 \qquad\qquad (8\text{–}12)$$

where A_R = attenuation due to Rayleigh scattering at a wavelength λ
    (dB/km)
    C_0 = constant $C_0 = 0.7$ (dB/km) $\times \mu m^4$
    λ = wavelength (μm).

This equation applies to multi-mode fibers. The attenuation due to Rayleigh scattering is generally smaller for single-mode fibers. The total intrinsic attenuation for silica fibers is the sum of these three components. See Figure 8–5.

The curve exhibits a minimum at about 1.6 μm, typical for all silica or glass fibers. At lower wavelengths, Rayleigh scattering is predominant, and at longer wavelengths infrared absorption increases rapidly.

FIGURE 8–5   Losses in silica fibers.

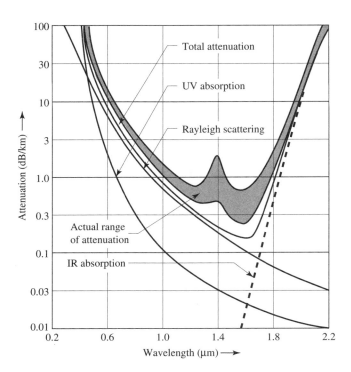

Besides the mechanisms described above, **extrinsic factors,** such as impurities in the materials, manufacturing techniques, and methods of fiber deployment, affect the fiber attenuation. The most significant impurity is the water radical OH$^-$ that causes an attenuation peak at 1.38 μm. The magnitude of the peak depends on the concentration of impurities and is typically 0.5 to 2 dB/km. Other factors are lightguide imperfections, such as material defects and stresses introduced during the manufacturing process. A deployment-dependent loss factor results from bends or microbends in the fiber that cause distortion of the core and increase losses. As a result, the actual attenuation curve is somewhat different from the one shown in Figure 8–5. The actual curve more closely resembles the shaded area of the graph. This curve demonstrates that the attenuation in real fibers shows two distinctive minima, often called the **windows,** for best propagation of the signal. For silica fibers, one window is about 1.3 μm, the other about 1.6 μm.

8–3   CLASSIFICATION OF FIBERS

Classification by fiber bandwidth, attenuation, and numerical aperture is the most meaningful way to categorize fibers. These characteristics depend on fiber type (as described) and fiber material. By far the most common fiber material is silica and its numerous variants. Plastic and plastic cladded fibers are also produced.

The plastic fibers are mainly composed of polymethylmethacrylate (known as PMMA) and polystyrene. Their refractive indexes are 1.49 and 1.59, respectively. The first, PMMA, can be used as cladding for a polystyrene core. Fluorocarbon polymer or a silicone resin can also be used as core material.

Plastic fibers are inexpensive and easy to use. They can be fashioned into large diameter fibers, up to 1 mm, that are easy to handle using simple connectors. They can be cut with a razor blade and, because of their flexibility, can be bent around tight corners. Since they are step index fibers, they have a large numeric aperture on the order of 0.5.

The disadvantage of plastic fibers is their very high losses. They exhibit several sharp C-H resonances and have several windows in the 500 to 700 nm region, the visible range for which inexpensive LED sources are available. The attenuation in the windows is high, typically about 100 to 200 dB/km. The dispersion is usually 200 to 600 ns/km, corresponding to a bit rate of less than 5 MB/skm. Also, the refractive index of plastics is quite temperature sensitive, limiting the upper usable temperature of the fiber to below 100°C.

The characteristics of **plastic-clad silica fibers** fall in between the all-silica fibers and plastic fibers. They have two windows at 800 and 1310 nm with an attenuation of about 5 to 10 dB/km. Since they can be produced only as step index fibers, their dispersion is high.

Despite these shortcomings, plastic fibers are excellent for applications where limited bit rates are required over short distances—for example, the interconnection of computers in a building. Their cost and ease of installation give them a distinct advantage in such installations. Table 8–1 gives a summary of fiber characteristics and typical applications.

The specific strength of silica fiber is very close to that of steel. Since the fibers have a small diameter, the absolute strength of a fiber is small and it must be protected from external forces. The fiber surface must be guarded from scratches and the entire fiber length must be saved from sharp bends that increase fiber attenuation. Thus, the fiber has to be enclosed in a protective envelope or made into a fiber cable. The construction of the cable depends on the fiber protection required. The cost and construction requirements of a submarine cable that has to withstand pressure of $100 \ GN/m^2$, be protected from marine organisms, and remain in service for decades are different from these of a cable that simply connects several computers in the same building.

Environmental conditions that the fiber must be protected from include:

1. **Surface scratches.** These allow the flux to escape from the fiber. A plastic coating, often called the **primary coating,** is used to protect the cladding surface.
2. **Longitudinal stress.** Including steel members in the cable and configuring the fibers in a loose helical spiral that accommodates some elongation of the cable eliminates longitudinal stresses.

TABLE 8-1
Summary of Fiber Characteristics

Material	Type	Core/Cladding Diameter (μm)	NA	Attenuation (dB/km)	Bandwidth MB/skm	Typical Use
All plastic	multimode step index	200–600 450–1000	0.5–0.6	330–1000	low	very low cost, short haul (100 m)
Plastic cladding	multimode step index	50–100 125–150	0.2–0.3	4–15	4–15	low cost, short haul, low bandwidth
Silica	multimode step index	50–400 125–300	0.16–0.50	4–50	6–25	low cost, short haul, low bandwidth
Silica	multimode grad. index	30–60 100–150	0.2–0.3	2–10	150–2000	medium haul, medium bandwidth systems, lead laser systems
Silica	single mode step index	3–10 μm 50–125 μm	0.08–0.15	0.5–5	500–40,000	long haul, high bandwidth, laser systems
Silica	single mode soliton				up to 100,000,000	intercontinental wide bandwidth system

3. **Sharp bends.** These can be eliminated by stiffening the cable with sheathing or wire wraps and by installing the fiber loosely in the cable.

4. **Water infiltration.** Gels such as petroleum jelly, silicone greases, and high-viscosity polyisobutylene will block out water. Infiltration may be an acute problem with underwater cables that operate under very high water pressure.

5. **Rodents.** Buried cables require stainless steel sheathing to prevent gnawing by animals.

6. **Temperature.** Cable materials must have compatible expansion co-efficients so that no stresses are transferred to the fiber as the sheathing expands and contracts.

As Figure 8–6 shows, cables can be classified into three general groups:

1. Cables with minimum protection to be used indoors in a protected environment, such as in ducts [see Figure 8–6(a)].

2. Telecommunication cables to be used underground or hanging from poles. These require strength wires and rodent protection [see Figure 8–6(b)]. In these cables, the bundles of optical fibers are often loosely placed in the grooves of a plastic core. Many variations of such cables are available.

3. Special purpose cables. Military and submarine applications usually require cables with more protection [see Figure 8–6(c)].

8–4 FIBER OPTICS SPLICES, CONNECTORS, AND COUPLERS

A fiber optics link requires many components. Most common among them are splices, connectors, and couplers. Such components make connections between the fibers and their terminating devices. In low-frequency electrical communication, their effect on the quality of the link is minimal. In optical communication, they all introduce losses that must be considered. The principal requirement for a connector or splice is simple—it must direct all the flux from one fiber to the other. To achieve this transfer, the cross sections of the cores must be equal and core alignment must be nearly perfect. Considering the small diameter of fiber cores (a few micrometers to a few hundreds micrometers), these requirements are not easy to fulfill. Figure 8–7(a) shows the perfect condition for flux transfer and Figure 8–7(b), an actual fiber condition resulting in flux losses.

We can assign the following term to the coupling efficiency:

$$\eta_c = P_2/P_1 \qquad \qquad \textbf{(8–13)}$$

where η_c = coupling efficiency
P_1 = optical power (flux) before the joint (W)
P_2 = optical power (flux) after the joint (W).

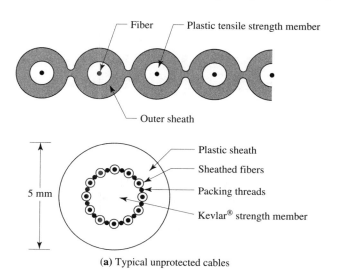

(a) Typical unprotected cables

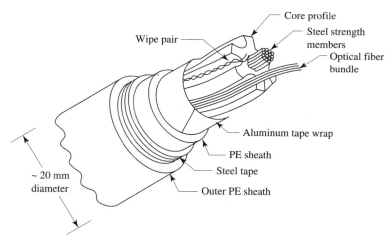

(b) Typical protected communication cable

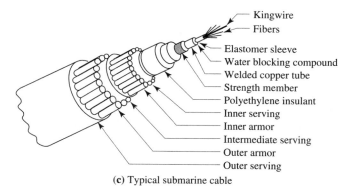

(c) Typical submarine cable

FIGURE 8–6   Samples of typical fiber optics cables.

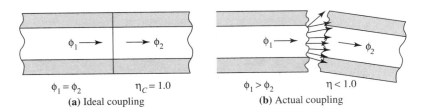

FIGURE 8–7   Ideal and actual couplings.

It should be noted that the custom in fiber optics is to use the term **power** (expressed in watts) instead of **flux.** The efficiency or loss of power in the coupling can also be expressed in decibels:

$$L_S = -10 \times \log \eta_c \qquad (8\text{–}14)$$

where L_S = coupling loss (dB).

Many factors affect the loss in a fiber connection. First, the propagation mode, single or multimode, has a considerable effect on losses. A second factor is the mode disturbance caused by the connection. The stabilized mode in the transmitting fiber is not necessarily transferred to the receiving one. New modes may be introduced that must travel before they achieve equilibrium. Thus, the loss depends on the distance from the joint where measurement is made. The third factor is the dimensional alignment, or misalignment, of the fibers. Last, the type of connection is a very important factor. Connections can be classified into two groups:

1. **Connectors.** In this category the fiber ends are mechanically aligned with a third medium, usually air, between the ends of the fibers.
2. **Splices.** Here, the ends of the fibers are fused together so that no medium with a different refractive index exists between the ends of the fibers. Instead of fusion, glues with optical characteristics similar to those of the fiber can be used for splices.

With so many parameters affecting coupling losses, it is difficult to give an accurate equation for every possible condition. Following is a set of equations that indicates the magnitude of losses under the most important conditions.

Losses in connectors with multi- or single-mode fibers are caused by the following conditions.

Fresnel loss is caused by a reflection at the air/fiber interface due to the different refractive indexes of air and the fiber (see Chapter 2), and can be calculated as follows:

$$\eta_F = [4n_1 n_0 / (n_1 + n_0)^2]^2 \qquad (8\text{–}15)$$

where η_F = Fresnel coupling efficiency

   n_1 = refractive index of the fiber core

   n_0 = refractive index of the media between the fiber ends. In the case of air, $n_0 = 1.00$.

and

$$L_F = -10 \times \log \eta_F \qquad \qquad \textbf{(8–16)}$$

EXAMPLE 8–5

Calculate the Fresnel efficiency and loss for a typical fiber with $n_1 = 1.50$ and air $n_0 = 1.00$ between the fiber ends.

Solution

From Equations 8–15 and 8–16, we have

$$\eta_F = [4n_1 n_0/(n_1 + n_2)^2]^2 = [4 \times 1.50/(1.50 + 1.00)^2]^2 = 0.922$$

$$L_F = -10 \times \log \eta_F = -10 \times \log 0.922 = 0.355 \ (\text{dB})$$

This loss is unavoidable and exists in all connectors with an air space between the fiber ends.

For calculating the other losses caused by inserting a connection into a fiber link, empirical equations do exist. They are reasonably accurate but involve many parameters that are difficult to measure. Who knows exactly the separation or face tilt of the fiber ends? So, instead of the equations, Figure 8–8 gives a set of graphs. From these graphs, the magnitude of the loss in a connection can be estimated.

As mentioned before, the loss in a connection does not depend only on the loss of flux in the connection but also on the disturbance of the **mode power distribution (MPD).** After traveling a certain distance, the wave forms a stable pattern or mode that stays the same as the wave travels along the fiber. When this pattern is disturbed, additional modes that have higher attenuation rates are introduced. The wave stabilizes again only after a certain distance of travel. A connection in a fiber introduces such a disturbance. Considering this, the loss in the connection is given under three conditions:

   A. Measured in a very short cable that is coupled into a wide intensity profile source, such as an LED.

   B. Measured at the end of a long fiber (about a kilometer) after the connection.

   C. Measured when the connection is between long fibers.

The graphs in Figure 8–8 clearly show the importance of fiber alignment in connectors. Considering that a typical multi-mode fiber has a diameter of 50 to 120 μm or about 2 to 5 mils, a lateral displacement of a few tenths

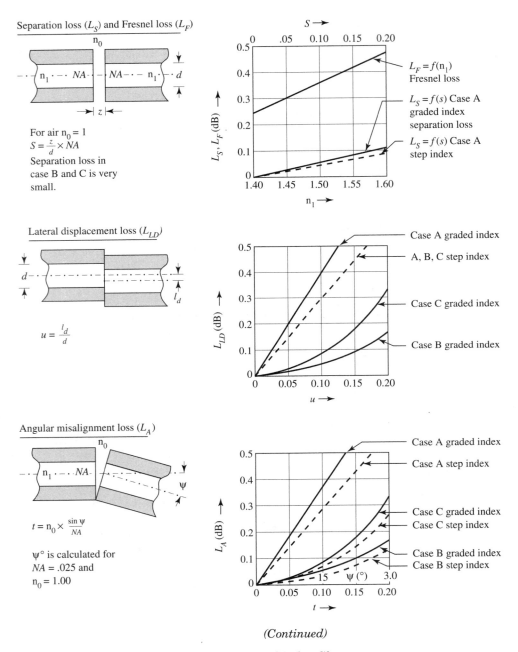

FIGURE 8-8   Connection losses in step and graded index fibers.

of a mil will cause considerable loss, which imposes severe requirements on the mechanical construction and precision of the connector. Besides precision

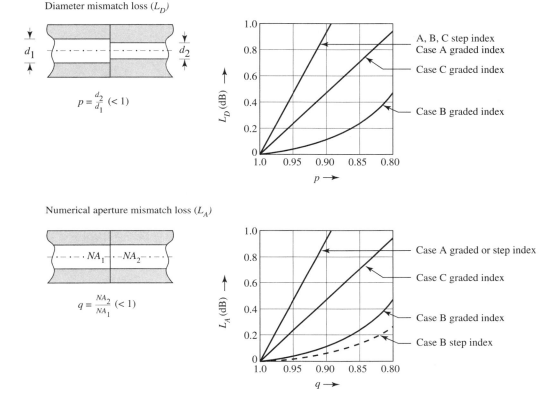

Diameter mismatch loss (L_D)

$$p = \frac{d_2}{d_1} \ (<1)$$

Numerical aperture mismatch loss (L_A)

$$q = \frac{NA_2}{NA_1} \ (<1)$$

FIGURE 8-8   *(Continued)*

alignment, ease of field assembly, reliability, environmental stability, and reasonable cost are required from a connector. These conditions are even more difficult to meet for a single-mode fiber where typical core diameter is about 10 μm.

A few basic connector alignment techniques are described in the material that follows. As Figure 8–9 shows, the connectors can be divided into three groups:

1. The fiber is squeezed into a precision V-shaped groove [see Figure 8–9(a)], which is a simple, low-cost technique. The shortcomings are that the alignment depends on the uniformity of the cladding and the bare fiber may be deformed in handling.
2. A precision alignment plug and guiding ferrule are used [see Figure 8–9(b)]. This type of connector is more robust, but more expensive since it requires very accurate mechanical parts.
3. The fiber is aligned using expanded and collimated beam connectors [see Figure 8–9(c)]. These connectors have a lens that expands the

beam to a wider diameter, reducing critical lateral displacement tolerances. Precision angular alignment is still required. Additional optical losses are introduced by the reflections and absorptions in the lenses. These connectors are typically used in industrial and military environments.

In practice, many variations of basic connector types exist. Multiple connectors allow ribbon cables to be stacked and connected using the precision groove method.

Splices are permanent joints in a cable. In a splice, the ends of the fibers are permanently connected by fusion, welding, or gluing. Thus, Fresnel loss is eliminated. The main loss in a splice is caused by lateral displacement of the fiber ends. Typical splicing loss in 10 μm diameter single-mode and 50 μm multi-mode fiber is shown in Figure 8–10. The alignment of the fiber ends in a splice is most important, as is clean and perpendicular fiber-end cutting. Figure 8–11 shows typical alignment techniques.

Figure 8–11(a) shows a **capillary splice,** where the fiber is inserted into a glass or ceramic capillary tube whose inner diameter is only slightly

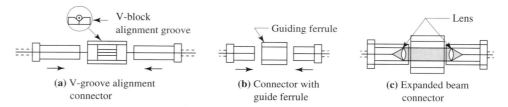

(a) V-groove alignment connector

(b) Connector with guide ferrule

(c) Expanded beam connector

FIGURE 8–9   Basic connector types.

FIGURE 8–10   Splice loss from displacement of fibers.

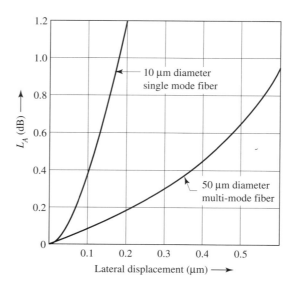

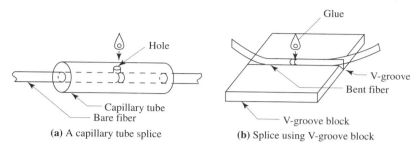

(a) A capillary tube splice   (b) Splice using V-groove block

FIGURE 8–11   Basic cemented splice techniques.

larger than the fiber diameter. A transparent glue (epoxy) is applied through
a hole in the capillary. Figure 8–11(b) shows a V-groove splicing block. The
fiber is forced into the groove by bending it. Then a drop of glue is applied to
join the ends. Several variations of this technique exist.

Fusion splicing joins fibers permanently by fusion or welding. With this
technique, a nearly perfect joint can be formed. However, the process is most
complex and requires skill and special equipment. An apparatus for fusion is
shown in Figure 8–12. This fusion-splicing device consists of two V-groove
alignment blocks. One block can be manipulated by a micromanipulator to
achieve a nearly perfect alignment of the fiber ends. The process is observed
through a microscope. Two electrodes use an electric discharge to fuse the
fibers. Lasers and microflame may also be used. Before fusing, the fiber ends
are cleaved using the same arc or heating device.

The process is more complex for single-mode fibers because the cladding
irregularity makes core alignment difficult. In this case, as optical power is
transmitted through the fibers, they are manipulated until maximum transmission, and therefore proper alignment is achieved.

Directional couplers are devices that couple the signal from one fiber
to another. They are used for building a network—distributing the same signal to several subscribers, inserting many signals into one fiber, or building
a bidirectional optical link. Some possible coupler configurations are shown
in Figure 8–13.

FIGURE 8–12   Basic fusion-
splicing device.

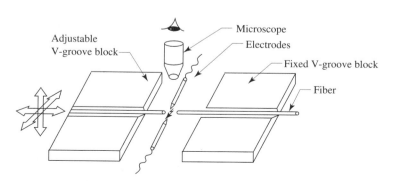

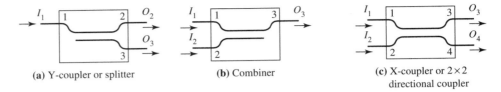

(a) Y-coupler or splitter   (b) Combiner   (c) X-coupler or 2×2 directional coupler

FIGURE 8–13   Basic optical couplers.

Figure 8–13(a) shows a **Y-coupler** or **optical splitter.** In a Y-coupler, power to input port 1 divides between output ports 2 and 3. When the power division between the outputs is not equal, the device is called a **T-coupler** or **optical tap.** The lesser output port can be used to monitor the power in the main line. Figure 8–13(b) shows an **optical combiner** where the signals from two input ports 1 and 2 combine into one output port 3. The **X** or **2 × 2 coupler** in Figure 8–13(c) performs both splitting and combining. Here, the signals from the two input ports 1 and 2 divide between two output ports 3 and 4.

Besides the schematics shown, many other configurations are possible. For example, the **N × M star coupler** where the signals from N input ports divide between M output ports and the **1 × M tree coupler** where the signal from one input port divides between M output ports. Optical multiplexers (MUX) and demultiplexers (DEMUX) are also available.

An ideal optical coupler is unidirectional; that is, no power transfers between the input ports and all input power transfers from input ports to output ports. In actual couplers, some signal is transferred between input ports and some power is lost in the transfer from input to output. These characteristics are specified using the following terms.

Directivity measures the isolation between the input ports. It is defined thus:

$$D_{ij} = -10 \times \log(P_j/P_i) \qquad (8\text{–}17)$$

where D_{ij} = directivity or isolation of input ports i and j (dB)
P_i = power applied to input port i (W)
P_j = power measured at input port j (W).

During this measurement, all other ports are terminated but no power is applied.

Power loss in the coupler is expressed as **transmission efficiency** and is defined thus:

$$\eta_T = \sum P_O / \sum P_I \qquad (8\text{–}18)$$

where η_T = transmission efficiency
P_O = total power measured from all output ports (W)
P_I = total power applied to all input ports (W).

Transmission efficiency can also be expressed as coupling a loss:

$$L_C = -10 \times \log \eta_T \qquad\qquad \textbf{(8–19)}$$

where L_C = coupling loss (dB).

EXAMPLE 8–6

Input port I_1 of a 4 × 4 star coupler feeds 250 μW optical power to the coupler. Total output power measured at ports O_4 to O_8 is 205 μW. At the input port I_2, 1.50 μW power is measured. What is the isolation between ports 1 and 2 and transmission efficiency of the coupler?

Solution
From Equation 8–17 the insulation between ports 1 and 2 is

$$D_{12} = -10 \times \log(P_2/P_1) = -10 \times \log(1.50/205) = 21.3 \text{ dB}$$

The transmission efficiency from Equation 8–18 is

$$\eta_T = \sum P_O / \sum P_I = 205/250 = 0.82$$

and the coupling loss from Equation 8–19:

$$L_C = -10 \times \log \eta_T = -10 \times \log 0.82 = 0.86 \text{ dB}$$

There are many ways to make a coupler. The simplest ones are shown in Figure 8–14.

The easiest way to make a coupler is to fuse two fibers together so that the flux from one fiber can escape to the other fiber, as in Figure 8–14(a). Several fibers can be twisted and fused together forming a 4 × 4 star coupler, as in Figure 8–14(b). Another method is to terminate the fibers in a transparent mixing section where flux from input ports is directed to the output ports. This principle is shown in Figure 8–14(c). Couplers are also formed on glass and semiconductor substrate. Much more sophisticated coupling devices, called **wavelength-division multiplexers (WDM)**, can insert or extract signals of specific frequency bands to the transmission links. These devices, together with transmitters, receivers, and amplifiers, are basic building blocks of fiber optics networks.

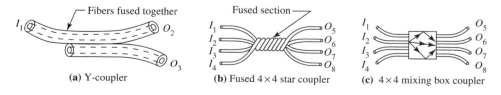

(a) Y-coupler   **(b)** Fused 4×4 star coupler   **(c)** 4×4 mixing box coupler

FIGURE 8–14   Simple optical coupler construction.

8–5   FIBER OPTICS SOURCES AND DETECTORS

Fiber optic sources must meet several specific requirements. For silicone fibers, the source frequency must be at a low loss window of approximately 850, 1300, or 1650 nm. The source intensity profile must be suitable for maximum flux coupling into the fibers. For a high data transmission rate, the source should have short rise and fall times. Cost, reliability, and aging effects are other considerations for selecting a fiber optics source.

LEDs and laser diodes meet these requirements. The LED is a simpler, less expensive device used for transmission rates of hundreds of megahertz. Laser diodes work for higher rates but are more expensive and complex, and are not as reliable as LEDs. Typical LED and laser diode characteristics are shown in Table 8–2.

In this text, the emphasis is on the LED, since it is the primary source for short length communication links. Table 8–3 lists **fiber optics LEDs** for fiber windows.

Two basic LED structures are available for fiber optics use: surface-emitting (SLED) and edge-emitting (ELED). In the **surface-emitting (SLED)** the flux is radiated perpendicular to the *p-n* layers and through the layers. These LEDs come in two types: edged-well and flat-surface, as shown in Figure 8–15.

TABLE 8–2
Comparison of LED and Laser Diode Characteristics

Parameter	LED	Laser Diode	Unit
Output power	1–10	1–100	mW
Power launched into fiber	0.0005–0.5	0.5–5	mW
Bandwidth at 800 nm	35–50	2–3	nm
Bandwidth at 1300 nm	70–100	3–5	nm
Rise time	2–50	<1	ns
Frequency response	<500	>500	MHz
Cost	low	high	

TABLE 8–3
Available LED Fiber Optics Sources

Material	Wavelength (nm)	Band Gap Energy (eV)
GaP	570	2.18
GaP/GaAsP	580–650	2.14–1.91
AlGaAs	650–900	1.91–1.38
GaAs	900	1.38
InGaAs	1000–1300	1.24–0.95
InaASP	900–1700	1.38–0.73

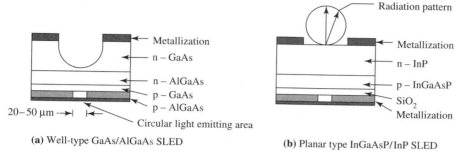

(a) Well-type GaAs/AlGaAs SLED

(b) Planar type InGaAsP/InP SLED

FIGURE 8–15   Typical SLED construction.

The radiation pattern of an SLED is close to a Lambertian pattern. Thus, SLEDs are difficult to couple into the narrow aperture of a fiber and, for that reason, are mostly used with multi-mode fibers having a higher numerical aperture. The fiber is sometimes fused to the well-type SLED for better coupling. This structure is called **pigtailed** construction. It requires a connector between the fiber and source. The disadvantage of the pigtailed LED is that its use is limited to the fiber of the pigtail.

The **edge-emitting (ELED)** emits radiation in a plane with the *p-n* layers, similar to the laser diode. As a result, its radiation pattern is unsymmetrical, or elliptical, as in a laser. Since the radiation angles in both directions are smaller than the SLED angle, an ELED is preferred with single-mode fibers. A typical ELED construction is shown in Figure 8–16.

A special version of ELED, called a **superluminescent diode (SLD),** is a cross between a laser diode and an ELED. In an SLD, light amplification through spontaneous emission takes place, but the diode does not have a feedback mechanism like a laser diode. As a result, higher intensities and narrower bandwidths can be achieved with an SLD.

In fiber optics communications, all calculations are based on optical power. The term **flux**, even though it represents the same quantity, is hardly

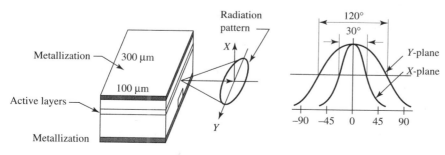

FIGURE 8–16   Construction and radiation pattern of an ELED.

used. The optical power generated in an LED can be calculated from the following equation:

$$P_{\text{LED}} = \eta_q i_{\text{D}} V_{\text{G}} \qquad (8\text{--}20)$$

where P_{LED} = optical power from the LED (W)

η_q = junction quantum efficiency (photons per electron)

i_{D} = LED current (A)

V_{G} = p-n junction energy gap (eV) (given in Table 8–3).

Since the quantum efficiency depends on the junction design, and is normally not known, this equation cannot be used for calculating LED output power. It points out, however, an important characteristic of the LED: optical power is proportional to the current through the LED.

Only a small fraction of the output power can be coupled into a fiber. The amount depends on the fiber type and the LED radiation profile. In the case of an SLED with Lambertian radiation profile and a step index fiber, the power into the fiber can be calculated from

$$P_{\text{F}} = P_{\text{LED}} T (NA)^2 \qquad (8\text{--}21)$$

where P_{F} = power coupled into the fiber (W)

T = transmission coefficient of the media between LED and fiber

NA = numerical aperture of the fiber.

The transmission coefficient considers the losses in the media and Fresnel losses at the surface of the fiber. As demonstrated in Example 8–7, the coupling losses can be considerable.

EXAMPLE 8–7

Calculate the power coupled from an SLED with $P_{\text{LED}} = 500$ μW into a step index fiber with $NA = 0.24$. The transmission coefficient between the LED and fiber is $T = 0.9$.

Solution

Using Equation 8–21, we obtain

$$P_{\text{F}} = P_{\text{LED}} T (NA)^2 = (500)(0.9)(0.24^2) = 25.9 \ \mu\text{W}$$

Only about 5% of the LED power is coupled into the fiber!

To improve the coupling between the source and fiber, special techniques are used to increase coupling efficiency. Often a lens, separate or etched into the LED, is used to narrow the output beam from the LED, as in Figure 8–17.

When using LEDs, thermal conditions and heat sinking, as described in Chapter 4, should be considered to achieve long life and reliability.

FIGURE 8–17 Methods to increase SLED coupling to fiber.

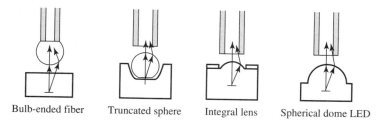

Bulb-ended fiber   Truncated sphere   Integral lens   Spherical dome LED

Communication bandwidth is one of the most important characteristics of a fiber optics link. Source rise and fall times are the factors in determining the bandwidth. Their response depends on the **carrier lifetime** (τ) at the junction and the type of LED used. The response of an LED can be enhanced by special drive circuits, often called **peaking circuits.** The peaking principle is shown in Figure 8–18. The LED driving waveform is distorted, forcing faster transfer during the turn ON and turn OFF period.

Basic circuits are used for driving LEDs—series and shunt switching. Shunt switching is considered somewhat superior to series switching. Typical LED driving circuits, with and without peaking, are shown in Figure 8–19, which also gives design formulas for the circuits. In series switching, a small trickle current is often fed through the LED to avoid zero bias operation of the LED, improving the turn ON time.

Properly designed peaking can improve the rise time and bandwidth by a factor of at least two. For greater bandwidth or a faster pulse rate, laser diodes should be used. Laser diodes are used for fast data rate and long haul transmission lines.

Fiber optics detectors are, almost without exception, photodiodes, described in detail in Chapter 6. Some of that material is repeated here and presented from the perspective of fiber optics, placing special emphasis on power and noise relationships.

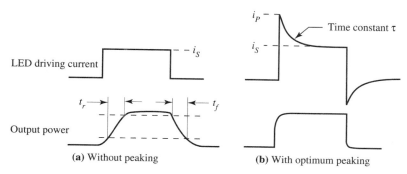

FIGURE 8–18 Peaked waveform for enhanced LED response.

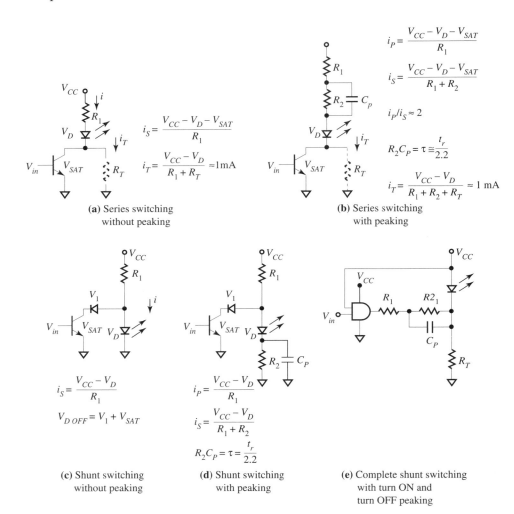

FIGURE 8–19 Principal LED driving circuits.

In fiber optics most calculations are based on optical power expressed in dBm. *The reference level of dBm is 1 mW.* Thus, the power in reference to dBm can be expressed as

$$\text{dBm}_1 = 10 \times \log(P_1/P_0) \quad \text{or} \quad P_1 = P_0 10^{(\text{dBm}1/10)} \tag{8–22}$$

where dBm_1 = power level of P_1 expressed in dBm

P_1 = power level P_1 (mW)

P_0 = reference power level P_0 = 1 mW.

EXAMPLE 8–8

Express power levels of $P_1 = 4.7$ mW and $P_2 = 120$ μW in dBm.

Solution

$$dBm_1 = 10 \times \log(4.70/1.00) = 6.72 \text{ dBm}$$
$$dBm_2 = 10 \times \log(0.120/1.00) = -9.21 \text{ dBm}$$

A junction photodiode is a constant current source. Its merit can be expressed in diode **responsivity,** which is either the ratio of diode current or voltage across the load resistance to the power applied. Based on the simplified diode equivalent circuit, as shown in Figure 8–20(a), the current responsivity can be expressed as

$$RE_i = i_D/P = (\eta e)(\lambda \times 10^{-9})/hc = (0.804 \times 10^{-3})(\eta\lambda) \qquad \textbf{(8–23)}$$

and

$$RE_v = RE_i R_L \qquad \textbf{(8–24)}$$

where RE_i = current responsivity (A/W)
RE_v = voltage responsivity (V/A)
  P = applied optical power (W)
 i_D = diode photocurrent (A)
  η = quantum efficiency of the junction (electrons/photons)
  λ = wavelength (nm)
  e = electron charge 1.60×10^{-19} (C)
  h = Planck's constant 6.63×10^{-34} (Js)
  c = velocity of light 3.00×10^8 (m/s)
 R_L = load resistance (Ω).

Quantum efficiency depends on the semiconductor material and junction construction. For Si and InGaAs, it is about 0.80; for Ge, about 0.55 at the wavelength of peak response.

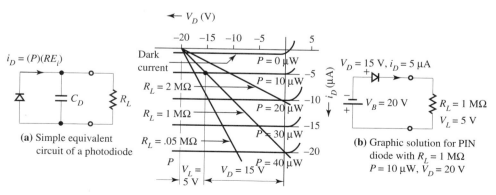

(a) Simple equivalent circuit of a photodiode

(b) Graphic solution for PIN diode with $R_L = 1$ MΩ $P = 10$ μW, $V_D = 20$ V

FIGURE 8–20   PIN diode equivalent circuit and characteristics.

EXAMPLE 8–9

Calculate current responsivity for the following semiconductor materials at their peak response wavelength.

    Silicon 800 nm
    Germanium 1550 nm
    InGaAs 1700 nm

Solution

Using Equation 8–23, we obtain

for silicon $RE_i = (0.804 \times 10^{-3})\eta\lambda = (0.804 \times 10^{-3})(0.80 \times 800) = 0.51$ A/W
germanium $RE_i = (0.804 \times 10^{-3})(0.55 \times 1550) = 0.69$ A/W
InGaAs $RE_i = (0.804 \times 10^{-3})(0.80 \times 1700) = 1.09$ A/W

Typical PIN photodiode i_D/v_D characteristics are shown in Figure 8–20(b). As a rule, the diode is always used with reverse bias, which reduces the junction capacitance and dark current. The output voltage from the circuit is

$$V_L = (P)(RE_i R_L) \tag{8-25}$$

where V_L = voltage across the load resistor (V)
    R_L = value of the load resistor (Ω).

Using Kirchhoffs' Law gives the voltage across the photodiode:

$$V_D = V_B - V_L \tag{8-26}$$

where V_D = voltage across the photodiode (V)
    V_B = bias voltage (V).

In the case illustrated in Figure 8–20(b), the bias voltage is $V_B = 20$ V, the output voltage $V_L = 5$ V, and diode voltage is $V_D = 15$ V. The diode has linear response as long as the diode stays in the reverse bias condition; that is, $V_D \leq 0$, which corresponds to maximum optical power

$$P_{max} = V_B/RE_i R_L \tag{8-27}$$

where P_{max} = maximum power for diode linear region (W).

In the case of this example, $RE_I = 0.5$ A/W, $R_L = 1$ MΩ and the maximum power is $P_{max} = 20/(0.5 \times 10^6) = 40$ μW, which is the applied power at the crossing point of the load line with the i_D-axes. Two more load lines are drawn on the diode characteristic: $R_L = 2$ MΩ and $R_L = 0.5$ MΩ. As the load resistance increases, the load voltage also increases. At the same time, however, the linear power range decreases.

The rise time or frequency response of the circuit is determined by the rise time and junction capacitance of the diode and the value of the load resistance. The rise time of the diode is determined by its construction and is given in the diode data sheet. For PIN diodes used in optic communication links, rise time is typically 1 ns.

The time constant formed by the diode capacitance and load resistance causes a circuit rise time; accordingly

$$t_C = 2.19 \times 10^{-12} R_L C_D \qquad (8\text{--}28)$$

where t_C = circuit rise time (s)
C_D = diode capacitance (pF)
R_L = load resistance (Ω).

The combined rise times of the diode and circuit rise time can be calculated:

$$t_R = \sqrt{t_D^2 + t_C^2} \qquad (8\text{--}29)$$

where t_R = receiver time constant (s)
t_D = diode rise time (s).

Rise time limits the system bandwidth. The 3-dB point of the bandwidth can be calculated from the rise time, using

$$f_R = 0.35/t_R \qquad (8\text{--}30)$$

where f_R = receiver circuit 3-dB bandwidth (Hz).

EXAMPLE 8–10

For a receiver circuit with PIN diode rise time of $t_D = 2$ ns and capacitance of $C_D = 5$ pF calculate:

a. receiver 3-dB bandwidth when $R_L = 1$ MΩ
b. value of R_L when 10 Mz system bandwidth is required.

Solution (a)
From Equation 8–28, we get

$$t_C = 2.19 \times 10^{-12} R_L C_D = (2.19 \times 10^{-12})(10^6 \times 5.00)$$
$$= 10.95 \times 10^{-6} \text{ s} = 10.95 \ \mu\text{s}.$$

Since $t_C \gg t_D$, it is obvious that t_C determines the bandwidth.
From Equation 8–28:

$$f_R = 0.35/t_C = 0.35/10.95 \times 10^{-6} = 0.032 \text{ MHz}$$

The bandwidth is very narrow because of the high $R_C C_D$ time constant.

Solution (b)
From Equation 8–30, the receiver time constant for 10 MHz bandwidth should be

$$t_R = 0.35/f_R = 0.35/10 \times 10^6 = 0.035 \ \mu\text{s} = 35 \text{ ns}.$$

Thus from Equation 8–29, we can write

$$t_C^2 = t_R^2 - t_D^2 = 35^2 - 2^2 = 1221 \text{ ns}^2$$

From here, $t_C = 34.9$ ns $= 34.9 \times 10^{-9}$ s $= 2.19 \times 10^{-12} R_L C_D$ or
$$R_L = 34.9 \times 10^{-9}/(2.19 \times 10^{-12}) C_D$$
$$= (34.9 \times 10^{-9})/(2.19 \times 10^{-12}) \times 5.00 = 3.19 \text{ k}\Omega$$

Here again the circuit time constant is the most important factor in determining bandwidth.

Example 8–10 shows that the circuit time constant plays a decisive roll in determining the bandwidth. A simple circuit, called a **current-to-voltage converter,** reduces the effect of diode capacitance (see Figure 8–21).

An operational amplifier is used in a current-to-voltage converter. The bias source and diode are directly connected to the inverted input of the amplifier. Since this input is at ground potential, the entire bias voltage is always across the diode. Because the amplifier input draws practically no current, all diode current is directed through the feedback resistor R_F. Thus, the output voltage is $V_0 = -i_D R_F$. This circuit has two advantages. First, the load line for the diode is practically zero resistance (a vertical line on the diode characteristics), so the circuit has a wide dynamic range. Second, the time constant of the circuit is not determined by the $C_D R_L$ time constant, but rather by the feedback resistor R_F and its stray capacitance C_F. Selecting a proper resistor or using several resistors in series improves the time constant and circuit bandwidth considerably.

Besides PIN diodes, **avalanche diodes** are widely used as fiber optics detectors. As explained in Chapter 6, avalanche diodes have an internal gain mechanism that increases their responsivity up to a factor of 100. They also exhibit a fast rise time, less than 1 ns. Thus, they have an advantage in low-power circuits where fast rise time is required, since they can deliver appreciable gain into a low load resistance. Their disadvantage is that they need a relatively high bias, on the order of 100 V.

8–6   A SAMPLE FIBER OPTICS DATA LINK

Now, having learned about fiber optics components, it is time to put together a fiber optics communication link. There are many other considerations in this process, such as modulation methods, detection methods, and noise

FIGURE 8–21   Current-to-voltage converter.

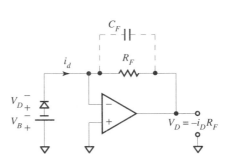

considerations, that are beyond the scope of this book. Therefore, we will design a very simple digital data link and show the design steps in this process. For more advanced designs, consult the reading list at the end of this chapter.

Three factors must be considered in the design of a data link: bandwidth, power levels, and error rates. Unfortunately, the parameters influencing one factor also affect the others. Consequently, no procedure guarantees that a solution for one factor is also satisfactory for the others. Therefore, backtracking and compromises must be made during design. There may also be several solutions for a given task.

The design sample considered here is for a digital data link to connect two computers in neighboring towns about 10 km apart. The baud rate for the computers is 30 Mb/s and the desired bit error rate (BER) is 10^{-12}. (The last term will be explained later.) The task is to select fiber, source, and receiver, and then calculate their power levels and error rates.

To begin, select components and then write **rise-time** and **power budgets** for the link. In the budgets, contributions from all components to the rise-time or power level are summed up so the link performance for these parameters can be evaluated.

The **rise-time budget** should start with determining the system or link requirements. Assuming that we use NRZ code, then the pulse duration and repetition period are equal: $T = 1/(\text{pulse rate})$; in our case $1/30 \times 10^{-6} = 33.3$ ns. To transmit this pulse, the system rise time should be about 70% of the pulse width or $33.3 \times 0.7 = 23.3$ ns. Thus, the total system rise time has to be $t_S \leq 23.3$ ns. This rise time is the sum of three components:

$$t_S{}^2 = t_{\text{source}}^2 + t_{\text{fiber}}^2 + t_{\text{det}}^2$$

Now we select our components.

Source: LED 820 nm GaAsAl emitter; couples 12 μW or -19 dBm into 50 μm diameter fiber; rise-and fall time 11 ns

Fiber: step index 50-μm core diameter glass fiber $NA = 0.24$; 5.0 dB/km loss; dispersion 1.0 ns/km

Detector: PIN photodiode, $RE^* = 0.38$ A/W; $NA = 0.40$; diode capacitance $C_D = 1.5$ pF; rise-time $t_R = 3.5$ ns; dark current $i_{DD} = 10$ pA

Next we can assemble the rise-time budget

Total system rise time (t_S)	23.3 ns
Source rise time (t_{source})	11.0 ns
Fiber rise time (t_F)	$10.0 \times 1.0 = 10.0$ ns
Allowance for detector rise time:	

$$t_{\text{det}} = \sqrt{t_S{}^2 - t_{\text{source}}^2 - t_F{}^2} = 17.9 \text{ ns}$$

RMS sum of all rise times $\sqrt{11.0^2 + 10.0^2 + 17.9^2} = 23.3$ ns

---

*Measured at the fiber mount.

Now calculate the receiver diode load resistance. Using just a load resistor for the detector diode, the time constant allocated for $R_L C_D$ is

$$t_C = \sqrt{t_{det}^2 - t_R^2} = \sqrt{17.9^2 - 3.5^2} = 17.5 \text{ ns}$$

From Equation 8–28, we can write

$$R_L = t_C/(2.19 \times 10^{-12})C_D = (17.5 \times 10^{-9})/(2.19 \times 10^{-12})(1.50)$$

$$= 5.32 \text{ k}\Omega$$

The **power budget** considers the source power and losses in the communication link. Assume we have four connectors with a 1.0 dB loss per connector; thus

Launched power into the fiber	−19.0 dBm
System losses:	
Fiber attenuation; 5.0 dB/km × 10 km = 50.0	dB
Four connectors at 1.0 dB; 4 × 1.0 = 4.00	dB
Total loss	54.0 dB
Power available at the detector	−73.0 dBm

Thus, the available power to the detector is

$$P_S = 1.00 \times 10^{-7.3} = 5.01 \times 10^{-8} \text{ mW} = 50.1 \text{ pW}$$

and with detector responsivity $RE = 0.38$, we obtain

$$i_D = P_S RE = (50.1 \times 10^{-12})(0.38) = 19.0 \text{ pA}$$

and the photodiode output voltage is

$$V_L = i_D R_L = (19.0 \times 10^{-12})(5.32 \times 10^3) = 101.1 \text{ pV}$$

Error rate analyses determine the quality of the detector output signal by comparing it to the system noise. The output signal from the detector always contains a noise component that is superimposed over the average value of the signal. A typical condition is shown in Figure 8–22. It shows typical high and low signal levels in a digital system. Between these two levels is a decision level.

When the signal is above this level, it is registered as HI; below, it is LOW. In a practical system, the signal level is superimposed with a noise component that is a randomly varying waveform with an amplitude distribution as shown at the right side of the drawing. The noise signal consists of a variety of amplitudes, with the higher peaks having a lower probability of occurrence. There is always a possibility that the LO level noise has a peak, as shown at time A, that reaches above the decision level and may be registered as HI. Also, the HI level may have an opposing noise peak that may register LO, as at time B. Thus, an error in transmission may occur.

When the noise amplitude and signal amplitude are known, the probability of error occurrence may be calculated. The result is given as the **bit-error rate (BER)** and is expressed as a fractional quantity. A BER = 0.01

FIGURE 8–22   Bit error caused by random noise.

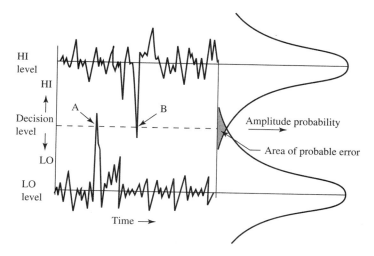

means that in every 100 bits transmitted there may be one bit in error. The lower the BER, the better the quality of transmission.

A typical acceptable BER is 10^{-9} for telecommunication systems and 10^{-11} to 10^{-12} for computer data transmissions. A BER rate 10^{-9} means that 10^9 bits can be transmitted with a probable error of one bit. In our case, we transmit 30×10^6 bits per second, so probable error occurs once in every $10^9/30 \times 10^6 = 33.3$ seconds. This is a short time, but it does not mean that only messages shorter than 33 seconds can be transmitted. All digital systems have error-checking methods (parity check, for example) that detect the errors and retransmit the signal. A system with a high BER is therefore less efficient.

The main noise sources in fiber systems are thermal noise and shot noise. The equations for them, given in Chapter 6, are repeated here.

$$i_{Trms} = \sqrt{4kT\Delta f/R_L} \tag{8–31}$$

$$i_{Srms} = \sqrt{2e(i_D + i_{DD})\Delta f} \tag{8–32}$$

where i_{Trms} = RMS value of thermal noise current (A)
$\quad i_{Srms}$ = RMS value of shot noise current (A)
$\quad\quad k$ = Boltzmann's constant; $k = 1.38 \times 10^{-23}$ (J/K)
$\quad\quad T$ = absolute temperature (K)
$\quad\quad R_L$ = diode load resistance (Ω)
$\quad\quad \Delta f$ = system bandwidth (Hz)
$\quad\quad e$ = electron charge; $e = 1.60 \times 10^{-19}$ (C)
$\quad\quad i_D$ = diode signal current (A)
$\quad\quad i_{DD}$ = diode dark current (A).

In our case, $T = 300$ K and $\Delta f = 0.35/t_S = 0.35/23.3 \times 10^{-9} = 15.0$ MHz:

$$i_{Trms} = \sqrt{(4)(1.38 \times 10^{-23})(300)(15.0 \times 10^6)/5.32 \times 10^3} = 6.83 \text{ nA}$$

and thermal noise power of

$$P_{\text{TN}} = i^2_{\text{Trms}} R_{\text{L}} = (6.83^2 \times 10^{-18})(5.32 \times 10^3) = 0.248 \text{ pW}$$

The shot noise current is

$$i_{\text{Srms}} = \sqrt{2ei_{\text{D}} \Delta f} = \sqrt{(2)(1.60 \times 10^{-19})(19.0 + 10)(10^{-12})(15.0 \times 10^6)}$$

$$= 373 \text{ pA}$$

and the shot noise power is

$$P_{\text{SN}} = i^2_{\text{Srms}} R_{\text{L}} = (373^2 \times 10^{-24})(5.32 \times 10^3) = 0.74 \times 10^{-15} \text{ W}$$

In this case, $P_{\text{T}} >> P_{\text{S}}$ and the system is **thermal noise limited.** This condition is usual at lower power levels. At higher power levels, $P_{\text{S}} > P_{\text{T}}$ and the system is **shot noise limited** or **quantum limited.** The method for BER calculation for either system is different. As a rule, the quantum-limited noise gives a better BER than the thermal-noise limited system, but it is more complex to calculate than a thermal-noise limited system.

We will calculate the BER for a thermal-noise limited system since thermal noise is predominant. The basis for a thermal-noise limited BER calculation is the **signal-to-noise ratio (SNR)**. In our case,

$$S_{\text{SN}} = 10 \times \log(P_{\text{S}}/P_{\text{N}}) \tag{8-33}$$

where S_{SN} = signal-to-noise ratio (dB)
$\quad\quad P_{\text{S}}$ = signal power (W)
$\quad\quad P_{\text{N}}$ = noise power (W).

In our case,

$$S_{\text{SN}} = 10 \times \log(50.1 \times 10^{-12}/0.248 \times 10^{-12}) = 23.1 \text{ dB}$$

The BER rate for a thermal-noise limited system can be read from the probability error chart depicted in Figure 8–23.

This chart shows that up to an SNR of 15 dB, the error rate improves very little. For SNRs above 15 dB, the slope of the chart changes and a small improvement in SNR results in a considerable improvement in BER. For a BER = 10^{-12}, an SNR of 23 dB is needed. Our system just meets this. In a good design, reserves for aging and temperature effects and component tolerances are needed. Thus, our system is somewhat marginal.

As mentioned, the sample calculations presented here are not a comprehensive presentation for optical fiber system design. Our purpose is to give an idea of the steps required in a design. Transmission links with different modulation methods and fiber optics networks require a much more detailed analysis. For these problems, the books in the reading list of this chapter are recommended.

FIGURE 8–23 Error probability in thermal noise limited system.

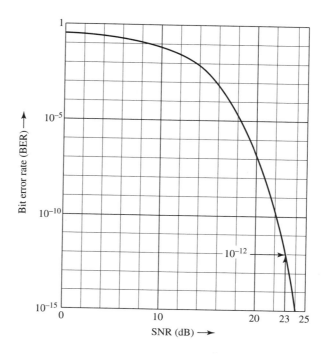

SUMMARY

Chapter 8 gives a short introduction to fiber optics fundamentals. Operating principles of optical fibers are discussed and main fiber types along with their characteristic bandwidths, dispersion, and losses, are described. Some fiber optics components (such as splices, connectors, and couplers) are described. A short review of fiber optics sources and detectors is presented and design principles for their use are given. Last, the design of fiber optics data link is presented and principles of error rate analyses are introduced.

RECOMMENDED READING

Palais, J.C. 1992. *Fiber Optic Communications*. Englewood Cliffs, NJ: Prentice-Hall.

Allard, F.C. 1990. *Fiber Optics Handbook for Engineers and Scientists*. New York: McGraw-Hill Publishing Co.

Goward, J. 1984. *Optical Communication Systems*. Englewood Cliffs, NJ: Prentice-Hall.

Kao, C.K. 1982. *Optical Fiber Systems*. New York: McGraw-Hill Book Company.

Hecht, J. 1993. *Understanding Fiber Optics*. Indianapolis, IN: Sams Publishing, a division of Prentice-Hall.

PROBLEMS

1. Find the NA and critical angle of polystyrene fiber with $n = 1.59$.
2. Find θ_A and NA of plastic cladded fiber with $n_1 = 1.59$ (core) and $n_2 = 155$ (cladding).
3. Calculate bit rate and signal bandwidth of a 0.5 km long multi-mode cladded fiber with $n_1 = 1.59$ and $n_2 = 1.55$.
4. Calculate Fresnel and separation (0.1 of fiber diameter) loss of a fiber with $n_1 = 1.59$. Compare the result with Figure 8–8 graph.
5. Using Figure 8–8, find connector losses in a short (case A) graded fiber link for
   a. lateral displacement loss when $d = 100$ μm and $l_d = 10$ μm
   b. angular displacement loss for a fiber with $NA = 0.29$ and $\Psi = 1.0°$
   c. diameter mismatch loss when $d_1 = 100$ μm and $d_2 = 90$ μm
   d. numerical aperture mismatch loss when $NA_1 = 0.29$, $NA_2 = 0.27$.
6. In an X coupler 200 μW is applied to input port 1 and 1.5 μW is measured at input port 2 and 140 μW at output port 1 and 40 μW at output port 2. What is the directivity, transmission efficiency, and coupling loss?
7. How much power is coupled from a 250 μW SLED to an $NA = 0.23$ multimode fiber when $T = 0.85$?
8. A diode detector with $RE_i = 0.5$ A/V is used with 5.0 kΩ load and 5.0 V bias. Find maximum power the diode can deliver and diode and load voltage when 0.5 mW is applied.
9. A PIN diode having 10 pF capacitance and 15 ns rise time is used in a system with 15 MHz bandwidth. Calculate the value of load resistance.
10. 100 μW is coupled to 10 MHz bandwidth PMMA plastic fiber which has an attenuation of 200 dB/km and has two connectors of 1 dB loss each. What is the maximum fiber length for 10^{-9} BER?

The answers are in Appendix C.

APPENDIX A

In this appendix the values of the relative sensitivity curve of the C.I.E. standard observer are given over the entire visible range of 380 to 770 nm in 10 nm steps. Use linear interpolation to find values between these steps. Also, a semi-logarithmic graph for the same values is shown. It can be used for fast, but modest, accuracy work of about 5%.

To convert the relative efficiency to spectral luminous efficiency in lumens per watt, multiply $v_{\lambda P}$ by 683 and $v_{\lambda S}$ by 1700.

Values of the Relative Sensitivity Curve of the C.I.E. Standard Observer

Wavelength nm	Photopic $v_{\lambda P}$	Scotopic $v_{\lambda S}$	Wavelength nm	Photopic $v_{\lambda P}$	Scotopic $v_{\lambda S}$
380	0.0000	0.0006	580	0.8700	0.1212
390	0.0001	0.0022	590	0.7570	0.0655
400	0.0004	0.0093	600	0.6310	0.0312
410	0.0012	0.0348	610	0.5030	0.0159
420	0.0040	0.0966	620	0.3810	0.0074
430	0.0116	0.1998	630	0.2650	0.0033
440	0.0230	0.3281	640	0.1750	0.0015
450	0.0380	0.4550	650	0.1070	0.0007
460	0.0600	0.5670	660	0.0610	0.0003
470	0.0910	0.6760	670	0.0320	0.0001
480	0.1390	0.7930	680	0.0170	0.0001
490	0.2080	0.9040	690	0.0082	0.0000
500	0.3230	0.9820	700	0.0041	0.0000
510	0.5030	0.9970	710	0.0021	0.0000
520	0.7100	0.9350	720	0.0010	0.0000
530	0.8620	0.8110	730	0.0005	0.0000
540	0.9540	0.6500	740	0.0003	0.0000
550	0.9950	0.4810	750	0.0001	0.0000
560	0.9950	0.3288	760	0.0001	0.0000
570	0.9520	0.2076	770	0.0000	0.0000

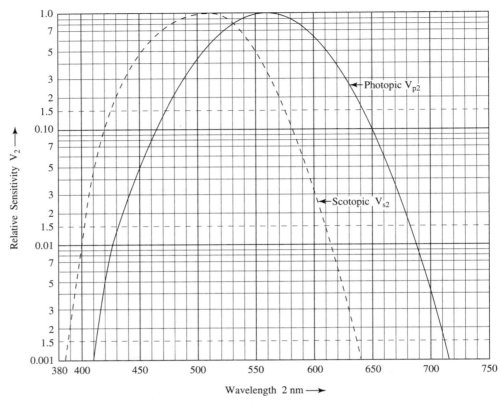

Graph of the relative sensitivity curves for the C.I.E. standard observer.

APPENDIX B

To compute flux from a given intensity profile is a time-consuming and arduous task. For those who have to do it often, a basic program is given in this appendix. The program is written for GW Basic and it works also with Q Basic.

Use of the program is illustrated in the following example:

EXAMPLE

Compute the flux from an intensity profile which is described by $I_\Theta = I_o \cos^n\Theta$ and $I_o = 2.00$ mcd. Compute the flux for two cases:

1. total flux over the range of 65 degrees using 13 slices, 5 degrees each, and
2. flux into 15-degree cone using 10 slices, 1.5 degrees each.

Compare the result with computations using equations from Table 1–3 and the result of numerical integration from Table 1–4.

Solution 1

Computing I_Θ for every slice and entering the data into the Basic program, we get the following printout:

```
Integration range: 65 deg.: 13 slices @ 5 deg.;
Center intensity: 2 millicandels.
Integration record:
```

Slice No. n	Degree nD	Intensity In	Omega n	Flux n
0.	0.0	1.0000	0.0240	0.0240
1.	5.0	0.9774	0.0479	0.0468
2.	10.0	0.9122	0.0955	0.0871
3.	15.0	0.8122	0.1423	0.1156

Slice No. n	Degree nD	Intensity In	Omega n	Flux n
4.	20.0	0.6885	0.1881	0.1295
5.	25.0	0.5542	0.2324	0.1288
6.	30.0	0.4219	0.2749	0.1160
7.	35.0	0.3021	0.3153	0.0953
8.	40.0	0.2021	0.3534	0.0714
9.	45.0	0.1250	0.3887	0.0486
10.	50.0	0.0705	0.4210	0.0297
11.	55.0	0.0356	0.4501	0.0160
12.	60.0	0.0156	0.4758	0.0074
13.	65.0	0.0057	0.2489	0.0014

The flux is   1.835 millilumens.

****** E N D O F T H E P R O G R A M. ******

Flux computed using equation $\phi = 2\pi I_o/(n + 1) = 1.795$ mlm.

The result from the example in Table 1–4 was 1.814 mlm. As seen, the program gives a slightly higher result. The accuracy of the program output improves when integration is performed using narrower slices, as shown in the next solution:

Solution 2

Again, computing I_λ for every slice and entering data, the printout is:

Integration range: 15 deg.: 10 slices @ 1.5 deg.;
Center intensity: 2 millicandels.
Integration record:

Slice No. n	Degree nD	Intensity In	Omega n	Flux n
0.	0.0	1.0000	0.0022	0.0022
1.	1.5	0.9979	0.0043	0.0043
2.	3.0	0.9918	0.0086	0.0086
3.	4.5	0.9816	0.0129	0.0127
4.	6.0	0.9676	0.0172	0.0167
5.	7.5	0.9498	0.0215	0.0205
6.	9.0	0.9284	0.0258	0.0240
7.	10.5	0.9036	0.0301	0.0272
8.	12.0	0.8758	0.0343	0.0300
9.	13.5	0.8453	0.0385	0.0326
10.	15.0	0.8122	0.0214	0.0173

The flux is   0.392 millilumens.

****** E N D O F T H E P R O G R A M. ******

The correct flux from the equation $\phi = 2\pi I_o(1 - \cos^{(n+1)}\Theta)/(n + 1) = 0.3868$ mlm. The program answer 0.390 mlm is closer to the true value compared to the previous case because of the narrower slices selected.

```
10 CLS
20 PRINT "    This program computes the flux from a point source for given"
30 PRINT " intensity profile. For this purpose divide the profile into N"
40 PRINT " slices with the width of D degrees each. The maximum number of"
50 PRINT " slices is 36. The program computes total flux from the source"
60 PRINT " when NxD covers entire range of the profile. Also, it may be"
70 PRINT " used to compute the flux into a narrower cone, with the cone "
80 PRINT " angle NxD."
90 PRINT "    To run the program select the type of profile and enter the"
100 PRINT " intensity values as prompted by the program."
110 PRINT "          ********** x ********** "
120 PRINT " DESCRIBE YOUR INTENSITY PROFILE. WHEN INTENSITY IS GIVEN IN
         CANDELS, "
130 PRINT " TYPE  cd; WHEN IN RELATIVE VALUE FROM 0 TO 1, TYPE  rel";
140 INPUT T$
150 PRINT
160 DIM I(36), PHI(36), OME(36), F(36)
170 INPUT "ENTER NUMBER OF SLICES"; N
180 INPUT "ENTER WIDTH OF SLICE IN DEGREES"; D
190 DR = N * D
200 PRINT "RANGE OF INTEGRATION IS"; DR; "DEGREES"
210 PRINT
220 INPUT "ENTER ABSOLUTE VALUE OF INTENSITY AT 0 DEG. Io = "; I00
230 INPUT "ENTER UNIT OF I0: cd OR mcd"; U$
240 PRINT
250 FOR X = 0 TO N STEP 1
260 PRINT "ENTER INTENSITY FOR SLICE "; X;
270 INPUT I(X)
280 NEXT X
290 PRINT
300 INPUT "IS THIS LIST CORRECT? yes/no"; C$: IF C$ = "no" THEN GOTO 320
         ELSE GOTO 360
310 INPUT "IS THE LIST CORRECT NOW? yes/no"; C$: IF C$ = "yes" THEN GOTO
         360
320 INPUT "TYPE THE SLICE NO. YOU LIKE TO CORRECT"; X
330 PRINT "ENTER NEW INTENSITY VALUE FOR SLICE"; X;
340 INPUT I(X): GOTO 310
350 PRINT
360 FOR X = 0 TO N
370 PHI(X) = X * D
380 OME(X) = .1097 * D * SIN(PHI(X) * .0175)
390 IF X = N THEN OME(X) = OME(X)/2
```

```
400 If X = 0 THEN OME (X) = .05485*D*SIN(D*.0175)
410 F(X) = I(X) * OME(X)
420 FTOT = FTOT + F(X)
430 NEXT X
440 PRINT
450 PRINT "Slice No.   Degree    Intensity   Omega n    Flux n
460 PRINT "   n           nD        In "
470 PRINT " -------------------------------------------------"
480 FOR X = 0 TO N
490 PRINT USING "_ ##."; X;
500 PRINT USING "_     ###.#"; PHI(X);
510 PRINT USING "_     ###.####"; I(X), OME(X), F(X)
520 NEXT X
530 PRINT
540 PRINT
550 PRINT "         The range of integration is "; PHI(N); "degrees; "
555 PRINT "         The flux is";
560 IF T$ = "cd" THEN PRINT USING "####.###"; FTOT ELSE PRINT USING
    "####.###"; I00*FTOT;
580 IF U$ = "cd" THEN PRINT " lumens." ELSE PRINT " millilumens."
590 PRINT
600 INPUT " DO YOU LIKE A PRINTOUT ? yes/no"; P$
610 IF P$ = "no" GOTO 810
620 LPRINT "         Integration range: "; N; " slices "; D; " degrees each;"
630 LPRINT "         Center intensity: ";I00;
640 IF U$ = "cd" THEN LPRINT " candelas." ELSE LPRINT " millicandelas."
650 LPRINT "         Integration record:"
660 LPRINT
670 LPRINT " Slice No.  Degree    Intensity   Omega n    Flux n"
680 LPRINT "   n          nD  "
690 LPRINT "-------------------------------------------------"
700 FOR X = 0 TO N
710 LPRINT USING "_ ##."; X;
720 LPRINT USING "_    ###.#"; PHI(X);
730 LPRINT USING "_    ###.####"; I(X), OME(X), F(X)
740 NEXT X
750 LPRINT
760 LPRINt "         The flux is ";
770 IF T$ = "cd" THEN LPRINT USING "####.###";FTOT ELSE LPRINT USING
    "####.###"; FTOT*I00;
780 IF U$ = "cd" THEN LPRINT " lumens." ELSE LPRINT " millilumens."
790 LPRINT
800 LPRINT "    ***** E N D  O F  T H E  P R O G R A M. *****": GOTO 830
810 PRINT
820 PRINT "         ***** E N D  O F  T H E  P R O G R A M. *****"
830 END
```

APPENDIX C
Answers to Problems

Chapter 1

1. $f_c = 4.72 \times 10^{14}$ Hz $= 472$ THz
   $\Delta f_b = 0.284 \times 10^{14}$ Hz $= 28.4$ THz
2. $\omega = 0.0123$ sr
3. Using $v_\lambda = 0.263$, $\phi_P = 1.80$ lm
4. Using $k = 17.0$ lm/w, $\phi_R = 54.1$ W, $P_{LOSS} = 5.9$ W
5. $I_P = 8,280$ cd
6. $I_{15} = 76.6$ mcd
7.a. OTF $= 0.0100$, $\phi_{REC} = 6.28$ mlm
7.b. OTF $= 0.0438$, $\phi_{REC} = 6.11$ mlm
8.a. $E_P = 17.9$ lx
8.b. $E_P = 16.0$ lx
9. $L = 5,000$ cd/m^2
10. $L = 0.748$ cd/m^2

Chapter 2

1. Using $R = 0.85$ reflection loss is 27.8%
2. $i = 4.16$ in, $I = 0.694$ in
3. $f = 11.11$ mm, $i = 33.33$ mm, $o = 16.67$ mm
4. $f = 187.5$ mm
5.a. $\phi = 0.00434$ mlm, OTF $= 0.00115$
5.b. $\phi = 0.158$ mlm, OTF $= 0.0419$
6. Both cases $I = 2.28$ lx

7.a. OTF = 0.090
7.b. OTF = 0.00040
8. i_D = 75 μA
9. o = 25 mm, i = 16.67 mm, i_D = 118 μA
10. *F-NUM* = 3.00

Chapter 3

1. λ = 489 nm
2. λ_{Si} = 1,138 nm, λ_{Ge} = 1,879 nm
3.a. A continuous spectrum source with color temperature of 7,000 K—typically daylight
3.b. 100% saturated mixture of blue and red. Not a spectral color
3.c. A pale green color of 520 nm
4.a. K = 19.0 lm/w, color temperature 3,000 K
4.b. R = 11.8 Ω, K = 16.2 lm/W, color temperature 2,900 K
4.c. Cost at normal condition 0.00625 \$/1,000 lmh cost at extended life 0.00764 \$/1,000 lmh
5. i_P = 4.17 A for 10 times inrush current
6. P_{avg} = 102 mW, t_{AMB} = 40°C
7.a. I_{avg} = 27.4 mcd
7.b. I_{avg} = 24.0 mcd
8. R_1 = 9.30 kΩ, R_2 = 400 Ω for 5 mA current, R_2 = 100 Ω for 10 mA current
9. V = 75 V
10.a. $V_A \approx$ 13 V
10.b. $V_A \approx$ 27 V
11. i = 3.92 nA
12. R_L = 1.00 kΩ

Chapter 4

1. E_R = 303 mW/mm^2
2. d_0 = 0.818 mm
3. d_{Z1} = 0.97 mm, d_{Z2} = 3.90 mm
4. Θ_i = 26.4 mrad, d_{0i} = 0.161 mm, d_{LENS} = 1.54 mm minimum
5. f = 124 mm, i = 124.5 mm
6. d_{0i} = 2.84 mm, Θ_i = 0.284 mrad, d_Z = 3.31 mm
7. Z_{R0} = 0.310 m, α_{12} = 40.1, d_{0i} = 20.1 mm
8. α = 0.00498, f = 2.49 m, i = 2.49 m
9.a. Laser diode because of cost and small size
9.b. He-Ne or ruby because of cost and visible wavelength
9.c. CO_2 because of power output
10.a. Use safety goggles and never look into primary or reflected laser beam

10.b. Terminate high power laser beam with absorbing termination
10.c. Wear skin protection

Chapter 5

1. $V_{avg} = 0.50$ V, $V_{RMS,ON} = 1.42$ V, $V_{RMS,OFF} = 0.474$ V
2. $H = 25.4$ mm $= 1.0$ in, selecting W/H ratio 0.7, $W = 17.8$ mm
3. At 100 lx illumination: $L_{LED} = 221$ cd/m^2, $L_B = 15.9$ cd/m^2, $C = 0.928$, $CR = 13.96$—acceptable ratios
   At 1000 lx illumination: $L_B = 159$ cd/m^2, $C = 0.24$, $CR = 1.40$—not acceptable ratios.
4. Selecting 1.00 mA current through the divider, the resistance values are $R_1 = 4.75$ kΩ, $R_2 = 0.20$ kΩ, $R_3 = 0.10$ kΩ, $R_4 = 0.20$ kΩ, $R_5 = 1.050$ kΩ
5. Selecting 1.00 mA current through the divider, the resistance values and voltages at nodes are:

dB Level (dBm)	Corresponding Voltage (V)	String Resistance (kΩ)
-4	0.488	0.488
-2	0.614	0.126
0	0.773	0.159
2	0.973	0.200
4	1.225	0.252
6	1.542	0.317
8	1.942	0.400
10	2.444	0.502
12	3.077	0.633
14	3.874	0.747
	$V_{CC} = 10.000$	6.126

6. $R = 81.3$ Ω
7. $f_{min} = 200$ Hz, $t = 5.00$ s, $t_P = 1.00$ s, $DF = 0.2$, $L_D = 10$ cd/m^2

8.

Character	Row Binary	Row Hex	Column Binary	Column Hex
4	0100	4	011	3
k	1011	B	100	4
Σ	1111	F	000	0
r	0010	2	111	7
ü	1010	A	001	1
π	1101	D	000	0

9.

Character	Row Binary	Row Hex	Column Binary	Column Hex
5	0101	5	0011	3
D	0100	4	0100	4
r	0010	2	0111	7
ä	0100	4	1000	8
?	1111	F	0011	3
Ω	1010	A	1110	E

10.

Address	Memory
XXX000	111110
XXX001	100000
XXX010	100000
XXX011	111000
XXX100	100000
XXX101	100000
XXX110	111110
XXX111	000000

Chapter 6

1. $I_N = 1.20 \times 10^{-14}$ A, $D = 1.00 \times 10^{13}$, $D^* = 2.28 \times 10^{13}$

2. $\lambda_{max} = 20{,}670$ nm $= 20.67\ \mu$m

3.a. $R_a = 35.9$ kΩ

3.b. $R_b = 202$ kΩ

4. $R_1 = 1.75$ kΩ, at $E_D = 1000$ lx $P_{CD} = 0.158$ kΩ, $i_{LED} = 4.45$ mA

5. $R_L = 3.00\ \Omega$, $V_L = 0.48$ V, $P_L = 76.8$ mW, $n_S = 14$ cells, $n_P = 4$ cells.

6. Using circuit similar to Figure 6–35(b), we obtain

Full-Scale Range fc	Full-Scale Range lx	Diode Current μA	Feedback Resistor MΩ
10	107.6	0.538	1.860
30	322.8	1.614	0.620
100	1076.0	5.380	0.186
300	3228.0	16.140	0.062

7. Using a circuit similar to Figure 6–35(a), we obtain

Full-Scale Range		Diode Current	Input Voltage	Gain	Feedback Resistor
fc	lx	μA	mV		kΩ
10	107.6	0.538	5.38	186.00	1,850.0
30	322.8	1.614	16.14	61.90	609.0
100	1076.0	5.380	53.80	18.60	176.0
300	3228.0	16.140	161.40	6.19	51.9

8. $E_{FS} = 107,600$ lx, $i_{D,FS} = 0.538$ mA, $V_{OL} = 0.445$ V, $A = 2.24$, $F_F = 1.24$ MΩ, at 1,500 fc the indication is 89% FS.

9. From Table 6–4

9.a. Ge or InAs diode

9.b. PIN or avalanche diode

9.c. Avalanche diode

9.d. PIN or avalanche diode

10. From Figure 4–48(b) an $R_L = 2$ kΩ produces a 5.0 V drop at 2 mW/cm^2 incidance. Use emitter follower circuit.

11. $\phi_G = 0.2$ W, $i_G = 0.06$ mA, required $i_D = 5.00$ mA, $\Delta V_G = 1.67$ V, $R_G = 27.8$ kΩ

12. $i_L = 5.00$ mA, $i_{B1} = 1.00$ μA, $\phi_{B1} = 3.33$ μW, $E = 33.3$ μW/cm^2

Chapter 7

1.a. $\eta_A = 0.36$

1.b. $d/D = 0.5$, from Figure 7–3 $\eta_A = 0.4 = 40\%$

2. $A_{ABE} = 250$ nm%, $A_{ACD} = 10,000$ nm%, $\eta_\lambda = 0.25 = 25\%$

3. 1/128 in = 0.198 mm, $F = 2.52$ lp/mm

4. $NA = 0.187$, from Figure 2–14 $T_1 = T_2 = 0.86$, OTF = 0.0259 $\phi_S = 314$ mW, $\phi_{DET} = 8.13$ mW

5.a. $E_D = 0.982$ lx, $i_D = 0.481$ μA

5.b. $M = 0.30$, using $T = 0.9$ $E_D = 13.1$ lx, $i_D = 6.55$ μA

6. $\Theta_S = 14.0°$, $\phi_S = 3.57$ mW, OTF = 0.162, $\phi_R = 0.578$ mW, $i_D = 0.289$ mA

7.a. $CTR = 0.667\%$

7.b. $CTR = 600\%$

8. $VTR_D = 2.4$, $VTR_C = 0.00133$, CMMR = 1,800, CMR = 65.1 dB

9. $R_F = 340$ Ω, $i_0 = 20.0$ mA, $R_L = 490$ Ω

10. For $R_0 = 1$ kΩ $i_F = 20$ mA and $R_{F1} = 0.425$ kΩ,
For $R_0 = 10$ kΩ $i_F = 0.7$ mA and $R_F = 12.1$ kΩ

Chapter 8

1. $\Theta_C = 39.0°$, $NA = 0.629$
2. $\Theta_A = 20.7°$, $NA = 0.354$
3. $B = 15.0$ MB/s, $\Delta f = 7.5$ MHz
4. $\eta_F = 0.899$, $L_F = 0.46$ dB, from the graph $L_F = 0.46$ dB
5.a. $u = 0.1$, $L_{LD} = 0.39$ dB
5.b. $t = 0.06$, $L_A = 0.22$ dB
5.c. $p = 0.90$, $L_D = 0.92$ dB
5.d. $q = 0.93$, $L_{LA} = 0.6$ dB
6. $D_{1,2} = 21.2$ dB, $\eta_T = 0.9$, $L_C = 0.46$ dB
7. $dBm_F = -19.5$ dBm
8. $P_{max} = 2.00$ mW, $V_L = 2.5$ V, $V_D = 2.5$ V
9. $t_R = 23.3$ ns, $t_C = 17.8$ ns, $R_L = 812$ Ω
10.

Signal level	-10.0 dBm
Noise level	-97.8 dBm
Gross margin	87.8 dBm
Margin for 10^{-9} BER	21.5 dBm
Transmission loss	66.3 dBm
Connector loss	2.0 dBm
Cable loss	64.3 dBm corresponds to 0.312 km of cable.

Index